ding dong

How Ring Went from *Shark Tank* Reject
to Everyone's Front Door

ding dong

How Ring Went from *Shark Tank* Reject to Everyone's Front Door

Jamie Siminoff
Founder of Ring

and ANDREW POSTMAN

Published by Your First Step, Inc.

Paperback ISBN 979-8-9934181-0-0

Hardcover ISBN 979-8-9934181-2-4

Ebook ISBN 979-8-9934181-1-7

Design and Production by Credible Ink

Printed in the United States of America

First Edition

To my father, Bruce—I wish you could have seen this.

Hope Amazon.com delivers to heaven,

they do everywhere else.

It is not the critic who counts; not the man who points out how the strong man stumbles, or where the doer of deeds could have done better. The credit belongs to the man who is actually in the arena, whose face is marred by dust and sweat and blood; who strives valiantly; who errs and comes short again and again, because there is no effort without error or shortcoming; but who does actually strive to do the deeds; who knows the great enthusiasms, the great devotions; who spends himself in a worthy cause; who at the best knows, in the end, the triumph of high achievement, and who at the worst, if he fails, at least fails while daring greatly, so that his place shall never be with those cold and timid souls who know neither victory nor defeat.

Theodore Roosevelt

Contents

CHAPTER 1

FAILURE IS AN OPTION

It was going to fail. *I* was going to fail. In front of eight million people.

Fuck.

Mark Dillon, my engineer, stood in front of me, spooked, soaked in sweat. It looked as if he'd just climbed, fully dressed, out of a swimming pool. We were moments away from taping the *Shark Tank* segment that would make us or break us. And our product didn't work.

"Mark?" I play-acted calm, faked a smile. The production crew across the brightly lit studio was rubbernecking. "What's going on?" I asked him.

"It's not working," he managed. If I looked half as stressed as he did, my face would be a meme on Twitter once this aired. "The first three didn't work," he said.

He meant doorbells. We had brought our entire inventory. Four of them.

"You'll get the last one to work, right?" I said. I remembered what our *Shark Tank* producer had told me weeks earlier, when I tried convincing her to let us submit a pre-recorded demo from my garage, a nice, controlled environment. We were struggling with reliability then, too. "Nope, has to be a live demo," she said, "hence the 'live.' We need to see it work, no exceptions." Then, in some twisted way she must have thought would relax me, she added, "Hey, it's TV! Failure can play better than success!"

Now, on a fall day in 2013, as I stood backstage on the *Shark Tank* set on the Sony studio lot in Culver City, California, the nightmare had become real. We were down to our final shot. The odds had shrunk, and there was nothing to suggest the last doorbell possessed some magical power that the first three didn't. My dream of someday making it was about to be torn apart by a bunch of sharks.

O

The product we were trying to demo—my latest invention—was called DoorBot, a wifi-enabled doorbell that sent a live video feed to your smartphone or tablet. Someone rings your doorbell and you can see and talk to them, whether you're upstairs in bed, in the garage, or a thousand miles away. When I installed the prototype on the front of our house, my wife, Erin, threw me a curveball.

"This makes me feel safe," she said. She said it was my best invention yet.

Seriously? I didn't think Erin would hate it but I hadn't expected her to love it. She'd seen a decade's worth of inventions and ideas from me and my occasional collaborators. Like Unsubscribe.com—exactly what it sounds like—which automatically dropped you from unwanted email lists. (I couldn't figure out how to monetize it but at least I made my investors their money back.) I did Simulscribe, renamed PhoneTag, which turns voicemails to text; you probably use some of its technology today. (Did okay, not great, but at least I made my investors their money back.) I did PoketyPoke, later called Lasso, a service that alerts you when your conference call is starting. (No success, but at least I was never again late to a conference call.) I did Slow Down, Asshole!, a motion-activated, battery-operated camera and speed monitor positioned on front lawns to film passing cars; videos of those that speed—their IDs traceable via license plate—would be uploaded to a public web page. (Never got past the planning stage, did register the URL. A company called Flock more or less does this today. As I write, they're worth $7 billion.) I tried to bring inexpensive long-distance calling to eastern Europe and Africa via "Voice over Internet Protocol"—VoIP—but encountered challenges, like being chased by police on a Caribbean island, and barely missing a presidential assassination while installing a satellite receiver in Kinshasa. (Skype did something very similar; eBay bought them for $2.6 billion.) I did SNAP Garden—modular gardening tiles for the urban dwellers. (You've probably seen commercials for similar home-gardening products. Another success missed.) Erin was not thrilled with that one: My research and development involved pools of water on tables in our backyard, attracting hordes of mosquitoes.

Oh, and there was my first-ever invention, the summer after fourth grade, the perfect example that necessity *is* the mother of invention: We endured humid New Jersey summers without air-conditioning, so I jerry-rigged a blanket through which ice water was pumped. (A company called Eight Sleep does this today. Worth $1 billion.)

Do I sound like someone with ADHD? Who works hard and ends up with only singles or 500-foot foul balls? Who watches as other entrepreneurs get the most out of good ideas?

That might be true, all of it. But I do have one trait above all.

I never stop.

I'd built the doorbell prototype because with all that inventing going on in my garage-turned-lab behind our home in Pacific Palisades, a neighborhood at the west edge of Los Angeles, I was missing deliveries because I couldn't hear the doorbell ring. And my cell phone sometimes didn't get a signal. As with my fourth-grade cooling blanket, I was just trying to solve a problem I personally faced.

I assumed there was a video doorbell on the market already. I had just gotten an iPhone and figured I'd find something that worked on it.

I went to Best Buy. Home Depot. Staples. Amazon.com. Nothing.

How was I the first person with this problem? Come on, some enterprising inventor *must* have devised a solution... right? The closest I could find was a cheap battery-operated wireless doorbell with a speaker that plugged into an outlet in the house. The signal didn't even reach the garage.

As an inventor, I was always more self-taught MacGyver than trained engineer, so I built my own crude version of a wifi video doorbell. I bought an off-the-shelf wifi camera, made a housing for it with a 3D printer, removed our doorbell, plugged the camera into an outlet, and glued the whole ugly thing next to the front door. When someone pressed the button, I could "open" the camera on my iPhone, see my visitor on-screen, even talk with them. Problem solved. I was more excited about Erin letting me keep it on the house than I was by the possible ramifications of her enthusiastic reaction.

Another person who believed my doorbell was more than just a doorbell: my friend Diego Berdakin, a brilliant entrepreneur, USC professor, and the single smartest person I know. For months, he and I had been talking about developing a security system to protect the home, with motion sensors, door and window sensors, and an intercom with built-in cameras that would call the homeowner if something was amiss when they were away. Our working title was HouseNinja. We suspected it could do well, maybe really well, because (a) startups and bigger companies were just beginning to make "smart home" applications like Nest, which had gotten everyone's attention with its "smart" thermostat; and (b) who doesn't care about safety and security? The alarm company ADT was the Coca-Cola of the home-security industry. Diego and I both thought there were better ways to do things. We believed new technology would disrupt this multibillion-dollar market, as it had countless others.

Yet when I came up with my video doorbell out of necessity, I never imagined home security might start at the front door.

I wondered if I should shift my attention to just one thing, my DoorBot. Erin encouraged it. So did Diego. But could I turn my back on all the other product ideas, like SNAP Garden, or POP (a wireless phone charger), or Flo, a hydro-powered, wifi-enabled analyzer that shuts off the water supply if a leak is detected? They were all part of an "incubator" I'd named Edison Jr: Invest in Edison Jr. and you invested in every invention that came out of it. My first investor had been my brother, John. Investment from family automatically makes things more stressful, but it was also a nice confidence boost that he believed in me. My friend Jeff Natland, a great entrepreneur, also put in some seed money, as did Diego.

I hadn't raised enough money to bring all of my product ideas from concept to reality, so (ADHD alert) I came up with yet *another* idea to give my dreams a better chance of getting financed and turned into usable products and great companies. I started a crowdfunding, pre-sale platform for inventors, an alternative to Kickstarter, which had become the Kleenex of crowdfunding sites. I had a long list of ideas that would benefit from pre-sale, but Kickstarter banned inventors (as Verge.com put it) from posting "simulations or design renders to illustrate what a completed product may look like or how it may function." They allowed creators only to "provide photos or video of prototypes as they exist at the time of posting." So much for dreaming big. Kickstarter had narrowed whom and what it attracted, and largely became a place to fund and develop video games and movies. *My* platform would feel like a place for inventors, focused on products, not so much on software. I named it Christie Street, after the site of Thomas Edison's famous lab in Menlo Park, New Jersey, the first thoroughfare in America to get electric lights.

Now I needed a product example to launch Christie Street. After all, it's hard to have a pre-sale site without something to sell. I added DoorBot to the list.

My friend Loïc Le Meur, co-organizer of Le Web, Europe's largest tech conference and a perfect place to unveil Christie Street, invited me to speak there. Loïc was in Los Angeles a couple of weeks before the conference, so I invited him over to the garage to demo everything for him—POP, SNAP Garden, and now DoorBot.

"What do you think?" I asked Loïc when I was done pitching.

"Ze doorbell," he said without hesitation in his thick French accent.

"Really?"

"Ze doorbell," he repeated.

"But don't you—"

"*Doorbell!*"

On December 6, 2012, at Eurosites Les Docks in Saint-Denis, with a sea of Silicon Valley–adjacent players in the audience taking notes on their iPads, I pitched DoorBot. I showed a video I'd made with my friend Karni Baghdikian, which highlighted the product's "three pillars": convenience, monitoring, and security. We showed a mom answering the door on her phone while sitting in the backyard with her kids. We showed a UPS guy delivering a package. The one thing we *didn't* show: Did it make you feel safer at home? I worried that people would call bullshit on such a grand claim and lose any trust in the product. I concluded my presentation by announcing that DoorBot was the first product that would be available for pre-order on Christie Street today.

Afterward, everyone wanted to know more about my video doorbell. Nothing else.

What else can it do?

How long have you been working on it?

WHEN CAN I GET ONE FOR MY HOUSE???

My reaction to all this interest?

Fuck!

I'd just shown an auditorium full of really smart, better-resourced, better-connected people my apparently awesome new invention. I had almost no money to build it, test it, upgrade it, scale it, ship it. If it really was as cool and useful and needed as everyone kept telling me, no way the product wasn't getting stolen by someone who moments ago was just sitting out there in the dark, tapping away.

And just like when I came up with the wifi video doorbell idea to begin with, and assumed that Best Buy or Home Depot or Amazon.com would carry a solution to my little problem, the answer was the same.

Nope. No one ran with it.

O

Then, a miracle: Orders started coming in—actual money!—to the Christie Street crowdfunding platform. More than I expected. All for DoorBot. Proof of concept! More than $300,000 in under two months, well above our goal of $250k.

Shit!

This was real. Now we actually had to build it. The money wasn't nearly enough to address all the technical issues we would face. I knew that hardware products cost way more than software ones before you have a showable, saleable product. But the tribe had spoken.

I approached my friend Rami Rostami, a brilliant businessman I'd met years before when working on Simulscribe. Like me, Rami was not trained as an engineer but had taught himself, and impressively kept one step ahead every time a product line outdated the previous one. My first big technical issue for DoorBot was a battery problem. Most homes do not have wires for the doorbell. If you had a video doorbell and the camera was on all the time, your battery would drain within days. I needed the device to "sleep" until someone pressed the button, then wake up, then go back to sleep. Rami and his team had had experience with motion-detection sensors that activated to capture information, then shut down to save battery life. Always game to entertain new ideas and businesses, Rami agreed to help me *and* put a little investment in, too.

Mark Dillon, a software engineer with expertise in servers, the cloud, and implementation, worked on the doorbell from New York. John Modestine and August Cziment, two young guys working with me in my garage in Los Angeles, shifted their focus from home-irrigation systems that grew tomatoes and parsley on your terrace, and ways to catch speeding suburbanites, to DoorBot. Working even crazier hours than normal, we soon dubbed ourselves the "Siminoff Brothers," with my 4-year-old son Ollie a key part of the gang. He loved hanging out in the garage with the tools and contraptions, the tabletops of computer chips and wires, and the cool guys in T-shirts and Air Jordans. When we would work late on Sundays, which was most of the time, Erin, Ollie, the guys, and I would have pasta dinners at our home, where I would smoke fresh bread from Bay Cities in Santa Monica on the grill and sprinkle on truffle salt. During the days, Olga Iglesias, nanny to Ollie but also to the rest of us, made tacos more delicious than you thought possible. We were not well-funded but we were well-fed.

It was the camaraderie of a true startup. There was so much about hardware we didn't know. And maybe it's not quite true to say I had the guys *totally* focused on the doorbell. When Diego dropped by the garage, I showed him our progress on DoorBot. But I couldn't resist presenting some of the other products the Siminoff Brothers were moving forward on: John writing patents for SNAP Garden, August making progress on Flo, and my strangest idea yet, a service that comes to your home to replace your empty propane tank. I was convinced we could take on Big Propane.

Diego blew a fuse.

"Stop the bullshit!" he said. "You've got a bunch of C-minus ideas here! You can do C-minus ideas in your sleep! What's the thing worthy of your maniacal attention? That you'll want to build for the next ten years? What the hell, man?!"

He was right. The doorbell was the thing.

Rami and his team moved forward on the design, I cleaned it up a bit, but they were making little progress on the battery problem, so I told him I was taking the job back. No hard feelings; he'd gotten us off to a good start. Like me, he had too many things going on.

Mark and I made solid technical advances. We were building a battery-operated high-definition camera, something no one had done at scale. But I needed more engineers. My friend Laurence Hallier, an entrepreneur, pointed me to a firm in Utah that could help with that. I needed a factory, and another friend introduced me to one in Taiwan called Tatung, which built washers, dryers, dishwashers, and—maybe their most famous product—rice cookers. My friend Jim Hyman, whose housewares company had done business in Asia for years, pointed me to a customs broker and an insurance company.

I felt the pressure of the Christie Street pre-orders, especially since I'd gone through most of the money and there were still so many issues to resolve—video, audio, wifi, the chime, you name it. It was laughable that I had started a hardware-centric crowdfunding platform when Mark and I knew almost nothing about hardware. Yet here we were, projecting a "trust us, guys, we got this!" image. Strained friendships and relationships were piling up. We were all feeling it. I had fired and re-hired August once already, John pretty much daily.

As I struggled to address the known bugs in the product and unearth sources of new investment, a contact from my Unsubscribe days asked if I would do him a favor. A friend of his, a budding entrepreneur named Adam Winnick, was building a tech business and wanted to talk to other entrepreneurs. My buddy thought my startup background was perfect. *In what way?* I wondered. *Scattered? Clever ideas that are ultimately not monetizable, at least by me?* Driving to meet Adam for lunch in Santa Monica, I berated myself. *You're literally working out of a garage, your main product is a doorbell, it currently doesn't work, and you're going to give someone ELSE pointers?* What a stupid use of my time.

Adam turned out to be a great guy with a pretty decent idea, and I shared some nuggets of experience and a few names. He was grateful, and I felt a little more upbeat that I had provided something approaching value. I knew how important it was to keep hustling and putting yourself out there, no matter your current situation. You never know.

Toward the end of the meal, Adam asked, "So what are you working on? Kevin says it's pretty cool."

I hesitated. I was almost too fed up to talk about DoorBot. I was 36 years old, having a professional midlife crisis, trying to keep my chin up. But I felt like I was failing as an inventor, and I hadn't had the big triumph I wanted, something that so many friends and acquaintances from the Silicon Beach community (LA's answer to Silicon Valley) had accomplished. Scott Marlette, my neighbor and close friend, who was employee #20 at Facebook. Sky Dayton, who had started EarthLink and so many others. My buddy Josh Kopelman, who had founded Half.com. My pal Jeff Natland, who co-founded Neteller (now Paysafe) and took it public *without* venture capital. The monthly poker game I'd been invited to was filled with huge successes, including Tinder founder Sean Rad, and I was just glad it was a $20 buy-in. They had all hit it big, some of them multiple times.

Even my favorite college professor, an expert on franchising and entrepreneurship named Steve Spinelli, had co-founded Jiffy Lube. I wanted my own dynamite success. I wanted to be *the guy*.

My dad, Bruce Siminoff, a great man and great father, hadn't wanted that for himself—or, for that matter, me. He lived a conservative personal and professional life and took few risks. The back half of his career he'd moved into mergers and acquisitions. Once, after laboring to put a deal together almost entirely by himself, he told me he'd taken a 5% stake in the company. He thought he was imparting sound business advice when he said, "James, you never want to own the whole thing. You don't want to be the guy. You want to be the guy behind the guy."

I was 16 years old and it was all I could do *not* to say, "Are you *kidding* me, Dad? I DREAM about being the guy!"

Now, two decades later, a garage inventor, I could say I *knew* people who had become *the guy*. I was adjacent to lots of them. I'd had dinners with them and gone kitesurfing with some. They made for great contacts and conversations. I was sure they were rooting for me. I was an inventor, and people—many of them rich, influential people—found me and my creative energy interesting to be around. I sometimes saw myself, or thought they saw me, as the court jester to their kings and queens. I was the Cosmo Kramer of my circle: the one with lots of ideas, many of them wacky, that never quite hit. The lack of success, for a middle-class kid from New Jersey, fueled me. I burned to be the one about whom everyone said, "He's the one who built *that*."

Sitting across from Adam, I could see he sincerely wanted to know about my progress. But even with the positive reception for DoorBot from Erin and Diego, the crowd at Le Web, and the people who had pre-ordered the product, I didn't see it as a home run. Maybe not even a triple. I quickly rehashed the origin story for Adam. *Couldn't hear doorbell from garage, searched without luck for one at Best Buy and Home Depot, I'm an inventor, had to build it myself, blah blah.* I ticked off the top technical challenges we continued to struggle with (inconsistent wifi, video quality, doorbell sometimes doesn't even ring) and the top business challenge (it was a *doorbell*, for starters, and hardware requires a shit-ton of money up front). I continued the litany for Adam. I'd hired some outside engineers to help, including a group in Salt Lake City. I'd committed to way more doorbell units than I needed or could afford, but the factory in Asia wouldn't manufacture them otherwise. Yeah, I had a few guys working crazy hours on it in my garage, and a little bit of money left from the pre-sale revenue to get us to the finish line. I conveniently left out that I was heading deep into the red, because even in my slump I knew that to get to the green, you had to hide the red.

"That sounds so cool," said Adam. "You should go on *Shark Tank*."

Yeah, no shit. What budding entrepreneur *wouldn't* want to go on *Shark Tank*, the hit reality TV show where entrepreneurs pitch a jury of highly successful capitalists itching to invest in promising startups?

Adam told me that *Shark Tank* had been reaching out more aggressively to the LA startup community, and he was on one of the emails. (Um, why wasn't *I*?) "They're looking for people whose business is a little bigger and further along than what they're getting," he said. "Higher-value products. Need to balance out the entrepreneurs baking cookies at home."

I knew enough about the show to be skeptical about the possibility. They received tens of thousands of applications a year, and a microscopic percentage got chosen. Of those who survived to make a video explaining themselves and their business, a bunch more got eliminated (entrepreneur boring, idea not as cool as it sounded on paper). Of those that made it through to get assigned to a producer, some got eliminated by studio executives. Of those that made it to taping, a bunch never aired, often because the exchange between entrepreneur and Sharks didn't work. Of those that aired, a bunch never got an offer. Sure, the exposure from being on the show would be great for business. *Could* be. I'd read that a 15-minute spot on *Shark Tank*, Friday night nationally at prime time, was equivalent to an $8 million commercial. But our product was barely ready. I didn't know when it would be. I really needed the money. Flubbing on air would be the final nail.

"Sure, I'll apply," I said. Of course I would—never stop hustling!—though I expected nothing.

Maybe sensing my lack of belief, Adam turned cheerleader. "You have a consumer product, it's electronics, it's real, it's got some complexity to it! It's unique. You've probably got IP. You're actually bringing in revenue, a real number. They're looking for things more like your stuff. Here. Email him."

He forwarded me the message from one of the *Shark Tank* producers. Nothing to lose. I emailed him from the table.

> Dear Shark Tank,
>
> We have a product, DoorBot, https://christiestreet. com/products/DoorBot, which might be interesting for the show.
>
> We are based in Santa Monica and would be happy to chat if you think this would be interesting.
>
> *Jamie*
>
> James Siminoff, Chief Inventor
> Edison Junior

Fifteen minutes later, on my drive back home on the Pacific Coast Highway, a *Shark Tank* producer called. "We gotta get you on the show, Jamie!" he said, as if it were up to me. "Your company's perfect!"

It was exciting to get such an immediate, enthusiastic response, but I had no illusions. The show sent me a very thick application with questions about our product and company, to be filled out in longhand and submitted with a brief video introducing me, my product, my company. Since my handwriting is close to unreadable, dictating my answers to August went late into the night. I had the option to mail it back or drop it off in person. One of my unbreakable rules: You're always better served connecting in person whenever possible, which explained why I had already accumulated a million miles flying, virtually all of them coach, many of them center seat. If there was even a small chance to meet face-to-face with someone who could help, I was there.

When I dropped off the completed *Shark Tank* form at a nondescript hotel in Sherman Oaks, the line of would-be entrepreneurs went on forever. The Hollywood expression "cattle call" suddenly made sense. Still, when I handed in the application, I made eye contact with a few of the team that mattered, and shook their hands, so they could put my face to a name. Days later, I got an email saying I'd been approved by ABC.

I kept it low-key in the email to my Siminoff Brothers:

> Boys we are going to be on mother fucking shark tank!!!!! Holy shit!!!!!!!

Shortly after, the *Shark Tank* producer assigned to me called to say how excited she was to work together, and how cool it would be if my segment featured a housefront with the doorbell on it.

"I love it!" I said. "I can't wait to see it!"

"No, *you* have to build it," she said. "And pay for it. Your decision. But I think that's the way to go."

I would tape on September 11. I couldn't shake the thought that the date wasn't exactly a sign of good luck.

There were still so many hurdles to getting an offer or even having our segment air. The *Shark Tank* producers were working with 200 other entrepreneurs—some individuals, some teams—and eventually had to whittle that group in half. I was going to prepare for the event like no one ever had. I did not want to be a joke. I wanted to be the best *Shark Tank* guest of all time.

I trained like I was an athlete preparing for the Olympics. I watched every single *Shark Tank* episode, pausing and slo-mo-ing at pivotal moments, noting where the vibe changed. I parsed the successful startups—Scrub Daddy, ChordBuddy, Nardo's Natural—and the unsuccessful ones. I studied the Sharks, their tells. Mark Cuban had made his fortune in tech and was often seen sitting courtside cheering like a maniac for the Dallas Mavericks, the NBA team he bought with his billions. As a tech guy, he was a great candidate for our business. Robert Herjavec was also tech (cybersecurity) but harder to read. Lori Greiner's background was QVC and inventing, and ours was a consumer product, so she was a good bet, too. Daymond John, fashion guy, probably not. Kevin O'Leary, sarcastically known as "Mr. Wonderful" for his unvarnished approach—the Simon Cowell of *Shark Tank*—also had a tech background, but I couldn't really guess him either.

In prepping, I wrote down hundreds of questions I might need to answer, based on what they had asked previous entrepreneurs. If they asked, "How did you come up with a valuation of seven million dollars?" I was ready: I wanted to be under a $1 million ask for 10% of the company; we were doing an annual run rate of $3 million; ~2x sales seemed valid and defensible. Oh, yeah: And prior to my *Shark Tank* season, only four companies had asked for a million or more, and none had closed a deal.

Our producer and I talked at least once a week. I found out later that some of the studio executives had wanted to cut me in a previous round. "It's a doorbell!" they said, to which she responded, "That's right, it's a doorbell! Nobody's reinvented it before! He did it to make his wife feel safe!" The executives finally gave in; each producer, as she said, "had one Get Out of Jail Free card." She'd used hers on me.

For the housefront that our producer had encouraged, I was fortunate to live in Los Angeles, where people build sets all the time, so I hired "Dr. Wood," renowned Hollywood set builder and carpenter, to build a plywood house façade for $8,000, a major hit to our remaining liquidity. (He'd built music video sets for Madonna and Ice Cube, plus sets for *In Living Color* and concert tours and commercials, so I was in good hands.) With the façade in our backyard, I set up chairs, just like the ones the five *Sharks* sat in, and recruited Erin and our neighbors to play Sharks.

"Be brutal," I told them. "Don't hold back."

I did my rehearsed pitch, got through the demo with no technical glitches, then they interrogated me. They couldn't stump me on the technology: While I was the founder and face of my company, my preferred title was Chief Inventor. I, along with my engineer, Mark, understood why every single piece in the DoorBot was there, where we got it, what it cost, what problems we had needed to solve, what made our device different from and better than every potential wifi camera maker and other competitor out there.

Questions about our core proposition would not be hard to answer either: I knew that people, whether they were investors, viewers, or consumers, needed to understand within 5 to 10 seconds, max, what the product did and its benefit to them personally. All I had to say was, "It's a camera for your front door that tells you someone's there. So you're 'always home.'" If an entrepreneur couldn't explain their purpose *that* simply and quickly, their business was probably DOA, like several *Shark Tank* entrepreneurs who'd pitched and been eviscerated on previous seasons. (One startup, called Attached Notes, featured a retractable board on your laptop monitor for your Post-it notes... *Why?*)

On the market side, I would play up the global "pre-awareness" boost that DoorBot enjoyed. Everyone knows what the traditional doorbell does and where it goes. There's no new paradigm to explain. Just show them how it's so much better than an existing doorbell.

The Sharks would push back, of course. Yes, we at DoorBot expected challenges. Once we got into retailers like Home Depot and Best Buy and Target, where exactly would they place us? In the few stores where you could buy one, the "Doorbell Section" was not exactly front and center; ironically, it was usually in the farthest place in the store from their own front door. And doorbells were not a fast-moving category: You replaced yours only when it broke, roughly once every couple of decades. The Sharks would surely question why shoppers would shell out $199, roughly 10x the average retail price, *for a doorbell.* I had an answer, sometimes five, for every question.

Mark Dillon had come out to LA for weeks from New York, away from his wife, working day and night to resolve the bugs in the product. John and August worked their usual insane hours. On September 10, the Siminoff Brothers loaded the tan-shingled, red-doored house façade into a U-Haul, which I drove to Sony Studios. The next day, Mark and I drove back to the studio in Erin's Jeep Grand Cherokee. He had tested the DoorBot back at the garage and it had worked perfectly, but he brought all four of our working prototypes, just in case. I was nervous but excited.

When we got to the soundstage, other entrepreneurs were pitching. We were told we'd set up right after lunch, following the entrepreneurs out there now, a company called Ten Thirty One Productions that staged live horror-themed events, like Haunted Hayrides and Great Horror Campout.

Fifteen minutes later I got bad news.

Ten Thirty One had just received a $2 million offer for a 20% stake in their company from Mark Cuban, the single biggest *Shark Tank* offer in the show's almost half-decade history, and almost three times larger than the previous record. Melissa Carbone, the company founder, accepted it.

Fuck. I was sure our own chances had just dropped way down. Certainly with Cuban, the Shark I'd been most optimistic about.

Focus. Don't think of anything but what a great product you have, people want it, you know how to sell it.

We had approximately 10 minutes to set up. The crew rolled the façade into place while Mark worked to get a DoorBot set up.

I chatted with our producer but couldn't help notice, out of the corner of my eye, that Mark kept scurrying from behind the façade to the doorbell on the front, back and forth. Each time he appeared, his shirt was drenched in another layer of sweat. I tried keeping my attention on our producer, but the next time Mark appeared he genuinely looked like a man who'd emerged from a sauna where he'd forgotten to take his clothes off. I smiled at our producer and excused myself.

I intercepted Mark. He said the first three doorbells we'd brought hadn't worked. He had no idea why.

We had built and tested everything a hundred times at my house, but that was on a wifi network unlike the one at the studio, with hundreds of devices spewing interference. We had even updated the DoorBots with new "firmware"—software embedded directly into the hardware—that we thought would help in "noisy" environments like this.

No such luck.

I knew how hard I'd been driving him, John, August, myself. And here we were. I angled my body so that our producer and the rest of the *Shark Tank* production crew could not see our expressions. I felt like punching something but couldn't show that I was about to lose it. There are times for screaming and punching holes in walls, and there are times for the soft touch.

I waved at the crew—*All good! Nothing to see here!*—but our producer was too shrewd not to smell something, and walked over to us. "Don't worry, you got this, Jamie!" She thought the lack of blood to my face was from regular nerves. "Just remember everything we practiced!"

I nodded, smiled insincerely, then led Mark by the elbow away from her. I could read his eyes. *Jamie's going to murder me. Legit murder, not sort of murder.* I thought of all the times I'd yelled at him, the fights, the calls to New York in the middle of the night to tell him something he'd done wasn't working quite right and could he stay up until it did? Now, his face glistening with perspiration, my lead engineer looked like he needed some genuine love. I leaned closer to him.

Calmly, I said, "Mark... make it fucking work."

Several moments later, which passed as if speeded up but also in super slo-mo, I found myself standing behind the double-door entrance to the *Shark Tank* set (an unusual way to start the pitch, our producer told me; that was good). I was terrified and ready. I had no idea if the fourth and final doorbell would work. I had no reason to believe it would.

Too late for certainty... *Action!*

I knocked. I could hear Mark Cuban on the other side of the doors say, "Who's there?"

"It's Jamie, here to pitch!"

I heard Daymond John say, "Who?"

"It's Jamie!"

Lori Greiner said, "Come in!"—and the doors finally swung open.

I strode forward. I was a barbell of confidence and terror. Seated across from me were Cuban, Herjavec, Greiner, John, and Mr. Wonderful. They looked eager.

"Sharks," I began, "wouldn't it be nice to know who's behind the door before you let me in?... "

I was off and running. I was trembling inside but tried my best to hide it. After introducing myself ("My name is Jamie Siminoff. I'm from Los Angeles, California"), I announced what I was looking for: "Seven hundred thousand dollars for a ten percent stake in my company." I talked about how billions were being spent on products that work with smartphones. I lightened the mood a bit with a fact about the doorbell. ("It was invented in 1880.") I pulled the DoorBot out of my back pocket, a dorky move I'd rehearsed endlessly for our producer. I tried hard to focus, but I kept realizing, *Holy shit, Jamie, you're on* Shark Tank*!*

Unfortunately, a second later, an inner voice countered with, *You're bleeding money. You agreed to pay for thousands more units than you have cash for. You really, really need a deal.*

It was time for me to ring the doorbell. Ring it and watch the image on my smartphone appear on the screen we'd set up on the housefront.

It had better show up crisp.

It had better show up, period.

I could feel my hand slippery with sweat, though I wondered if Mark, sitting behind the fake wall, had already drowned in his own perspiration. We were 0 for 3. We had one more shot.

I had my phone in my left hand. I pressed the button on doorbell #4 with my right.

One second passed.

And another.

CHAPTER 2

SAVE ME

The longest second of my life.

The image appeared.

The doorbell camera didn't just work. Miraculously, it produced the sharpest image on the screen we'd ever produced.

We were a go!

I dove into my well-rehearsed spiel for the Sharks, my nerves calming. They scribbled notes. Was Mark Cuban squirmy or was it just me? Maybe he was thinking about the $2 million he'd just offered to underwrite live-themed horror shows. I powered through. I talked about the added security DoorBot provided. "Think of it as caller ID for your front door," I said, which got a laugh from Lori.

I wrapped up my pitch with even more dork: I yanked a wad of money from my pocket, tossed the bills like confetti ("Own the dork!" our producer had said. "It's TV and we need this thing to stand out!"), and I challenged the Sharks, "Now, who wants to be the first to ring my bell?"

The Sharks looked engaged. Not too cynical. I started fielding questions, first from Lori. Yes, my product had two-way audio, one-way video. She wondered about suspicious people realizing that this was a camera lens and stepping away—or stealing the doorbell.

"Because all burglars ring the doorbell first," Mark interjected sarcastically, which got a howl from Lori.

Yes! I had the perfect answer! Erin's response showed me that the DoorBot was not just another gadget but something that could make you feel safer at home! Neighborhoods across the country were inundated by knock-knock burglaries and we had the answer!! Thank you, Mark, for the opening!

"You know what, though, Mark?" I said. "That's actually true! Burglars are not usually violent criminals. They *want* to see if someone's home. They want the opportunity to go into a house that does *not* have people in it. So it's actually true."

Mark nodded.

Yes! *Mark Cuban is going to invest! This is going to be good!*

They asked about the retail price. I told them $199, which got some sour looks, totally expected. They asked about the cost per unit. "Eighty-one-dollars and eighty-three cents," I reported. They wanted our sales history, and I told them we would be in our first retailer in two months—Staples. That we had launched nine months earlier and had an aggregate of $1 million in pre-sales, which impressed them. I told them we had just had our best month, $250k, and were growing month over month. So far there was no one competing directly with what we were doing.

This was tech. It was for the smartphone and the smart home. Change and disruption were coming. This wasn't about mere convenience but safety. Peace of mind.

Mark said he liked the product but wondered about bigger solutions that might obliterate us—

Lori interrupted. "I'm gonna jump in here. I think that you have potential to do a lot more with this... but I'm not connecting that this does enough at this time to distinguish itself as different from what else is out there on the market, for the higher price point. And so for that reason... "

Nooo. Wait, let me answer that—

She was a no. Floodgates, please don't open.

Mark Cuban had heard enough. "I gotta be able to say, you know what? When I jump in, I've got to add enough value that this company worth seven million could be worth eighty million, ninety million," he said. "I just don't see that progression. And for that reason, I'm out."

Herjavec echoed Cuban's skepticism, for his own reasons: He declined because it wasn't "an internet play but a consumer device," because hardwired doorbells had an advantage on us since they couldn't be hacked (Mr. Cybersecurity), and because he thought our price point was too high. Daymond John also passed quickly, which I'd expected.

There was one Shark left. No one was saying anything. *I need to get them talking.* I knew that without more discussion and exchanges, questions and answers, opportunity for tension or disbelief or humor—*drama! this is TV!*—there would not be enough material even to build a segment. I had been out there, start to finish, for not even five minutes.

Somehow, I got them talking again. I don't know how. Or they got me talking. A broader view on the product, the business, the market, the pitch, I think, but I have no idea what was said. I'm sure my voice was shaking.

Then it went quiet. The Sharks could smell blood in the water. I hoped my luck with the fourth and final DoorBot continued with the fifth and final Shark, Kevin "Mr. Wonderful" O'Leary.

He really dragged it out. This was TV, after all. "The only person left is Mr. Wonderful," he said. I did not think an opening in the third person was a particularly good sign. Jamie Siminoff would never do that.

But then Mr. Wonderful did what every entrepreneur on the show dreams a Shark will do: He made me an offer.

"I'll give you seven hundred thousand dollars, I want a ten percent royalty that drops down to seven after I recoup the seven hundred thousand, and I get five percent of the company's equity... today."

He was meeting my cash ask. He wanted *less* equity than I had offered. But he asked for a royalty, which I hadn't.

I did the only thing I could.

I refused.

O

Broke, I still had to refuse.

Mr. Wonderful was shocked. To be fair, *I* was shocked.

I was so desperate for money that my brain screamed *YES!*

Fortunately, my mouth said no.

Mr. Wonderful wondered why, almost angrily. He was prepared to give me the $700,000 I had asked for! But it was that 10% royalty... I just couldn't.

"We have a vision to build a big company out of this," I explained, "and I can't give someone ten percent of all of our sales because it will bleed us of cash when we really need it most."

Inside, I was dying. We needed the money bad. What we could do with $700,000!

But I also knew it was the right choice: I had prepped beforehand on what kind of offer I would say yes or no to. This was a bullet-in-the-head deal for the company. The royalty request would totally strip us of needed cash. (Had Mr. Wonderful asked for double the equity, I would absolutely have said yes.)

Most important: Turning him down meant my chance of airing had just plummeted. At least that's what I thought when the producers led me to the interview area.

They questioned me about what had just happened... *Did I think the Sharks had been fair? Unfair? What next?* They wanted the raw reaction. I felt as if I had sort of kind of been hit by a bus.

Moments later, out of the soundstage, in the bright California sun, all adrenaline gone, I stood there with Mark Dillon. I had let him down, along with Erin and the other Siminoff Brothers. I thought about Ollie. I knew he would never let me feel as if I had let him down. (Erin wouldn't either.) (And probably not the boys, either.) I'd just rushed through a nervous presentation, four Sharks had turned me down, the fifth made me an offer I couldn't accept, I had been on and off in 10 minutes, maybe 15 with the conversation I forced at the end. I had walked away from *Shark Tank* with nothing. I wondered what I was sure everyone who gets turned down wonders: *Will my segment even air? If it does, will it even matter?*

I called Erin. I told her I was in "shark shock." Yes, an offer was made. No, I had turned it down. I was grateful my wife had a steady job in production in the movie business. For me, it felt like an inevitable return to a familiar but uncomfortable place. Failed inventor.

Our producer appeared at my side. "My God, that was amazing!" she said.

"What?" I said, not comprehending. "That's never gonna air."

"No! You were amazing!"

"What are you talking about?"

"That was great TV! You killed it!"

"Really? I was on for ten minutes."

"Ten? You taped for almost an hour!"

Mark and I looked at each other, wondering whether we had lost our minds or she had.

"It felt like no time at all to me," I said.

I hoped she was right that my segment had worked well, because that meant a better chance at airing—but it needed to be soon. The show's season ran from October through May, and my segment could appear anytime in there, if at all. If it wasn't until spring, then the company would probably not make it to then. Even with the sales, we were quickly running out of money. I had product coming in soon that I had no cash to pay for. We needed the show to air no later than early December, around the time we'd be shipping for the holidays. That would help, a lot. The show's weekly ratings were phenomenal; like *American Idol*, it was a show that people talked about over meals. But that would matter only *if* the segment aired soon and *if* viewers got DoorBot's appeal. The Sharks certainly didn't seem to.

Mark Dillon packed up and took a car to the airport to fly home to New York. I called him from my car to tell him how grateful I was for everything we'd been through. *He'd* been through.

"Yep, thanks," he said. "That was the most stressful day of my life."

On the way home, in Erin's Jeep, I came close to crying. Thinking of the two constants in my life, Erin and Ollie, made it all sweeter and sadder. They'd been on this journey with me, as much as my investors or Mark or August or John or Diego or anyone else. I thought of how much comfort I got taking Ollie with me on business trips. We were like Batman and Robin. He had recently accompanied me to the factory in Asia where they were making thousands of DoorBots. I didn't need a Bring Your Kid to Work Day to bring my kid to work. At 5 years old, Ollie might have been the only person in that massive factory under the age of 20, but they had posted a sign at the entrance that read *"WELCOME JAMIE AND OLIVER SIMINOFF!"* which made me love them.

Back at the garage after the *Shark Tank* taping, John and August studied me. From their expressions I could tell they worried I might cry. Everyone had worked so hard to get to this point. I'd just turned down the one offer we got. I was mad for having blown over a month preparing for the show. We were in a worse place now than when I'd been invited. It had been a huge distraction and made a huge dent in our remaining resources. I was pissed at the Sharks for not getting it, all but Mr. Wonderful. Even though his offer was unworkable bordering on non-armed robbery, at least the guy believed enough to make an offer.

And I was mad at myself. For looking so down in front of the boys. As the leader, it was important for me to hide any stress I was feeling. I didn't want to betray even a tiny bit that the pressure was getting to me.

○

You're only as good as your people and I had good people. There was Mark Dillon, my engineer. (He was more or less CTO, though such a title at a company our size didn't make much sense.) He used to shuttle down to Los Angeles from San Francisco, until his wife took a job in New York and his commute got even longer. We'd met years earlier through his wife, one of my best friends in college, who said her husband was a terrific engineer. I invited Mark over to my house, and by the time he finished helping me build a fishpond, I hired him.

There was John Modestine, an industrial-design major I hired right out of college, Philadelphia University (now Thomas Jefferson University), to work on Edison Jr. projects, especially SNAP Garden. "Come to LA," I told him. "Work out of my garage for six months and see where it goes," code for "I might run out of money by then." John had never been west of the Mississippi. He showed up at LAX with two small bags and a big grin.

And there was August Cziment, a 22-year-old marketing major from the University of Colorado-Boulder. A couple of years earlier, when working at Unsubscribe, I had told a friend I was looking for someone "insane enough to work a hundred hours a week for almost no money, do anything and everything needed, and often take the blame and occasionally get fired for fuckups he had zero responsibility for. Though I'll always rehire him the next day."

"I think I know someone," said my friend.

Several nights later, August showed up at my house in jacket and tie, like he was interviewing at Morgan Stanley. It helped his case that my then 2-year-old son instantly loved him and dubbed him "Aju." I offered August a job. He surprised me by insisting on a formal offer letter before he would quit his job at a rental-car company. The day I handed him the letter, I told him, "Do me a favor and burn all those suits. I can't fucking stand suits."

He laughed, thinking it was a joke.

"Actually, forget that," I said. "Keep one suit for your wedding. Burn the rest."

That was my core team.

I scrambled to find us more money. I could write that sentence in any chapter of this book, about almost any month of DoorBot/Ring's existence. Right then, I needed a million bucks. Just before we taped *Shark Tank*, I had gotten the promise of a modest lifeline from First Round Capital, a venture capital firm co-founded by Josh Kopelman, a good friend with a ridiculous magic touch in his former life as entrepreneur. He had founded three very successful companies in a row: Infonautics, which went public in 1996; Half.com, which eBay bought in 2000 (and led them to launch the fixed-price "Buy It Now" format, which accounts for two-thirds of eBay's sales); and TurnTide, an anti-spam router business bought by Symantec. Having gone three for three, Josh said he knew the odds of his next startup being anything but a spectacular failure were overwhelmingly against him, so he did a mic drop and (his words) "went to the dark side and became a VC." Where, of course, his Midas touch continued: His firm, First Round Capital, was part of the seed round for Uber and Roblox, to name a couple. Josh K. and his partners had invested in dozens of entrepreneurs, including me and my partner Josh Roth, back when we did Unsubscribe.

When I pitched DoorBot to First Round, they found it... half-baked. They didn't love that my team and I had never shipped a high-quality hardware product. Hardware almost always requires multiple rounds of fundraising, meaning everyone's stake keeps getting diluted. It was a beast that needed continuous feeding. Would the pursuit be worth it in the end, whenever the end came? Would the equity that remained justify the years of endless weeks of countless hours of work, and their investment? And even if it did work out, the partners were wary that the market I was trying to break into was already getting crowded and at risk of competitors like Nest and Dropcam owning it. One partner wondered politely about our doorbell, "Who would ever put that piece of shit on the front of their house?"

But another partner, Chris Fralic, couldn't have been more pumped. He said, "I'm putting my badge on the table for this investment," and his enthusiasm and belief persuaded his colleagues to go in for a quarter-million dollars. Josh Kopelman wanted me to know that they were writing "a Jamie check rather than a DoorBot check," and wishing we were doing a different product. Because Josh and I were good friends, and he might find it tough to balance that with his fiduciary obligation, he asked Chris to oversee the investment. Chris was excited to do it.

The unwritten expectation (maybe it was written; oops) was for me to find other investors to fill out the rest of the investment "round." It's a common sequence with this type of fundraise: A bunch of investors want to come together; when they all do, you close the round. The risk profile for a company with $250,000 looking to raise $1 million varies greatly from one with a million in the bank. I had hoped that the $700,000 ask on *Shark Tank* would bring me to that golden threshold: With $50,000 promised to DoorBot by another investor, plus First Round's quarter-million, we would turn the *Shark Tank* success into a neat million, close the round, and proceed to the winner's circle.

But Mr. Wonderful's money wasn't to be. Nor was anyone else's. I needed to close First Round's part of the deal *without* the lead. I called the lawyer from First Round to have him prepare papers for their investment.

"We're good, we're gonna close the rest," I said, matter-of-factly.

There was a pause. "Okay," he said.

No double-checking? I hoped no one would confirm if what I said was closer to the truth or a lie. I desperately needed the money.

Within days, we signed the standard agreements, and First Round wired the quarter-million dollars to the DoorBot account. The next morning, Chris's name popped up on my phone.

"Jamie, what the fuck? Are we the only ones?"

"The rest of the money's coming, it's coming. I just needed to get going."

"Where's the other money coming from?"

"I've got a lot of good leads."

"So you're just bullshitting."

"We have lots of people interested. I just have to narrow it down to a few, then nail them down."

Another pause. "So we're the only ones," he said.

"Yeah. You're the only ones."

"That's great. Jesus Christ, Jamie."

"I really need the money, Chris."

"Okay," he said, calmly, and I will always love him for the way he said that one word. "You know that wasn't the intention."

"I know, I know."

Here was a guy who believed in DoorBot so much, he'd vowed to put his "badge on the table" for us. And in our very first inning of working together, I was already screwing him.

Another reason why this had to work, somehow.

○

I couldn't tell you the success rate of startups—2%? 0.2%?—but I know this: Your chance for success only goes up if you keep your head down and grind. I look at it like the lottery. The harder you work, the more lottery tickets you get. Work really, really hard, accumulate lots of tickets. Sure, even with thousands of tickets you're never guaranteed success. But everyone knows that your chance of winning the lottery is a hell of a lot better with a shitload of tickets than one.

DoorBots would soon be arriving from the factory in Asia. We would start filling the Christie Street pre-orders. And I signed a lease for a real office for us, in Santa Monica.

I scouted for more lead investors and got turned down by yet another one, a young former entrepreneur-turned-VC who rebuffed me because I didn't have "a strong lead." I responded to his rejection with a very cursory email, acknowledging its receipt.

But rather than keep it at that, or thank him for hearing my pitch (because you never know when you might need that person, or who knows who), I couldn't help myself. My dad always told me to never burn a bridge, but in the heat of the moment, I decided, what the hell, I would incinerate that bridge.

I followed up my first email to my latest rejector with a second one.

Just as a follow on to this, I hope that I did not come off as too much of a dick but I was really surprised by your email. Not because you passed but for the reason...

While a strong lead is great for a first time 20-year-old (maybe), it is really kind of a joke for a 36-year-old seasoned entrepreneur. I don't need a VC to help me with my A round...while I am happy to get help, what makes an A round happen is a fucking kickass company. Investors are whores, they invest in what makes them money, which makes the whole process pretty easy and efficient, especially at the A round+ levels.

Also, what really shocked me was I would expect an old-school asshat VC to tell me I need a lead at this point in the company, not a group of fellow successful empathetic entrepreneurs.

I really was excited to work with you guys and was really bummed when I got your mail. However, based on the reason, I would say it would probably not have been a fit anyway.

Thanks again and I am sure I will see you at something in NYC or SF in the near future,

Jamie

James Siminoff, Inventor

I was pretty pleased with the "investors are whores" line and calling an East Coast Ivy League elite bro an "old-school asshat." I pondered including the line "I was excited to work with rich New York white-shoe Harvard prick assholes only because I'm dying on the vine here," but thought better of it.

I'd like to say that my second email was impulsive, so I deserve some slack.

Yeah, no. I wrote it a full *week* after the guy turned me down. Fuck him.

The next morning, I stared at the message, wondering why I had sent it. I showed the email to Erin, like I was a little boy who knew he'd done something wrong.

"Did you really send that?" she asked.

"Yep, last night. Late."

She just shook her head and walked out of the room.

Here I was, in the trenches, and these untested VCs are lecturing *me* on what I really need? I had flown across the country to meet them in person. They made themselves out to be empathetic fellow startup people, and turned out they were just rich guys who'd probably gotten lucky with their investments so far, because they couldn't really, truly have understood what it was like. They didn't get it and maybe never would. They weren't actually in the arena, like the Teddy Roosevelt quotation at the beginning of this book, someone who errs and comes up short again and again, because there is no effort without erring or shortcoming.

I felt enough contrition for my nasty email that I forwarded it to Chris Fralic, who—I should probably mention—is the one who'd introduced me to the guy I'd just reamed. He wrote back:

> I don't normally see whore and ass-hat used in a
> single email like that.

And because few things get passed around more quickly than a juicy email, many in the VC world soon knew what I had written. I met Josh Kopelman for drinks two nights later and after his first sip, he said, "So... we're all whores?"

A couple of weeks later I finally got some good investor news, all thanks to our nanny Olga's tacos. Steve Temes, a New York VC now based in Miami, had pledged $50,000 for DoorBot, and was considering more. When he came through Los Angeles on other business, I invited him to the garage to see what we were spending his $50k on. He liked what he saw but made no sign that he was so blown away that he was ready to part with more money—until we sat down for lunch and he took a bite of Olga's taco.

"This is the best fucking taco I've ever had," he said, through a mouthful. He took another bite, swallowed, and said, "How much do you need?"

"Three hundred thousand," I said.

"Done," said Steve. "I love these damn tacos."

O

I landed a meeting with the actor Ashton Kutcher, who had shrewdly invested in various tech companies (Spotify, Uber, Shazam, Soundcloud, Airbnb). He invited me to his trailer on the Warner Brothers lot, where he was filming *Two and a Half Men*. While I had no trouble looking desperate in front of VCs, somehow I felt worse doing it in front of actors. He welcomed me into his trailer. The actress Mila Kunis, his girlfriend (and later wife), was sitting on the couch. "Why a doorbell?" she asked, as I was barely into my pitch. On one hand, I was pinching myself being there. On the other, I was just overwhelmed by the whole thing.

"How many doorbells have you sold?" she asked.

I told her. She, Ashton, and I had an energetic discussion about what this thing could possibly be in the future. I wished *they* had been Sharks.

But though they seemed to get it, the "investing in a doorbell" was a bridge too far. Ashton, with respect, made it clear that this was not an investment for him at this time. The trailer was small, and between that and rejection, I felt suddenly claustrophobic. Moments later I was out of there, standing in the sun on another Hollywood lot, taking big gulps of air. No new investment. Feeling shitty about our prospects. And it was my birthday.

My phone rang.

Shark Tank.

Someone was informing me that the DoorBot segment was going to air, in two weeks!

I guess I really had done well enough, somehow, to make the cut.

Before I could ask the caller any of my dozen questions, she'd politely hung up. I guess if your job is giving good news to people, some of whom are borderline desperate, you want to get in as many as possible. I owed Adam, my original *Shark Tank* contact, dinner at a restaurant of his choosing.

As I headed to my car, I wondered: Should I walk back to the trailer, knock on the door, and say, "Hey, Ashton? Um, Mila? By the way, we're going to pitch our product and company before millions of Americans, on one of the most popular TV shows around... want to invest *now*?"

I didn't. I'd had enough of no for a while that I wanted this beautiful yes to just sit there and linger, a perfect birthday present, unsullied by anything else.

The timing of the show was ideal—right before the holidays, with lots of units coming in that we could ship out, *if* we got more orders. Suddenly, overordering doorbells from the factory in Asia didn't freak me out so much.

I expected a big night. I contacted Tobias Lütke, CEO of an exciting startup called Shopify, an e-commerce platform built for way more traffic than we had ever gotten at GetDoorBot.com. They handled the kind of "flash sales" that could and probably would happen after my appearance on *Shark Tank* aired.

To say I was nervous—anal? insane?—is an understatement. On the patience meter, Tobi and I were clearly on the opposite ends of the spectrum.

> Hey James,
>
> Kickass to meet you. We are here for you - we've got tons of experience with flash sales (of proportions which are downright staggering) and TV appearances. There is nothing extra you have to do everything will just work.
>
> - tobi
>
> CEO Shopify

> Tobi,
>
> The show is Shark Tank and we are scheduled to air on November 15th. I believe it will start airing east coast at 9pm EST and then roll through the time zones. Shark Tank is notorious for taking down sites and I do not want to lose a single sale.
>
> Let's chat as I want to make 100% sure that we will not go down at all that night regardless of the traffic that comes in.
>
> Thanks!
>
> Jamie

DING DONG

James,

Being on Shark tank is a situation we are really familiar with. As you may know Mark and Daymond are both involved with Shopify through the modern incarnations of our Build a business contest...Part of the reason for their involvement is that Shopify is one of the few systems that survives their traffic bumps. They have been so impressed with that, that they have started shops to switch to Shopify before airing.

- tobi

CEO Shopify

Tobi,

Ok we are now 1 week away from airing on the 15th at 9pm

Please make 1,000% sure that we will not have a minute of downtime or any issues. I am hoping to do a ton of sales that night and would hate to miss even 1.

Jamie

No reply.

Tobi,

Just want to make sure we are all set for Friday

Jamie

James,

Our perf and ops team are aware but like I said, there is a Shopify store on Shark tank essentially every week. Your product is more compelling so it will be bigger than usual, but we can easily handle 10x to 100x the traffic (and have!).

- tobi

CEO Shopify

Tobi,

Yes I agree I think we might have more hits due to the more broad appeal of the product.

If 200,000 people went to the site at one time would it crash?

Due to the high price of the product we think we could do an insane amount of sales but are still very concerned about down time as we will never get those customers back.

Happy to pay to double whatever you guys are planning on having as infrastructure for us.

Jamie

James

We have 200k people hit the site every few seconds during normal operations. All of Shopify is one system and runs on over a million dollars of hardware. I understand your concern but... we got it :-)

- tobi

CEO Shopify

It's a miracle he didn't tell me to go fuck myself. I wish I'd had the foresight to invest in his company, since it's now valued at roughly $200 billion.

○

Two Fridays before Thanksgiving, the day my *Shark Tank* segment was airing, Erin and I called everyone we knew to come over for a viewing party.

Then again, she wasn't thrilled about having so many of our closest friends in our backyard, watching it unfold. Why? She was nervous that the show could "edit it any way they want, even make you look like a fool." I told her I didn't care. (Of *course* I cared.)

I am not a big drinker. I can go all in now and then, if the occasion is right—or wrong. My nerves got me as we set up for the party and I started drinking, "a little." But I soon saw that it was going to be okay, more than okay, because my brother, John, back in New Jersey, had seen the show, East Coast time, and called to tell us that I came off fine, no worries. He'd recorded the show and sent it to me. Erin and I sat in our living room and watched. My segment was edited in a way that actually made me come off pretty well. I did not look like an idiot (though I did look as if I was going to cry). The whole episode was cut down for time, of course, and they had not included some of my worst moments, in the interview area afterward. I breathed a sigh of relief, and kept drinking. It was like finding out you'd won a race two months after running it. I drank a little more. By the time our hundred nearest and dearest showed up, I was shitfaced.

I hung a white bedsheet tree to tree in the backyard so we had a big screen. Josh Roth brought his projector. So many of the people we loved most in the world had their eyes fixed on the bedsheet, and sometime after 8 p.m. Pacific, I simply smiled. At the taping two months earlier, I had thought one of the Sharks would save us. They didn't. Now I thought the airing could save us, and I realized how preposterous that sounded.

Save us. Come on.

As the rejections from the Sharks piled up on-screen, so many of my friends turned to me, sympathetic, agreeing that, yes, poor guy, it looked as if I was about to cry on national television. What sad eyes. The way I firmed and pursed my lips to keep my chin from shaking.

I just smiled, drunk and happy. It was the greatest 12 minutes of free advertising any startup could hope for.

CHAPTER 3

THE BEST WORST CHRISTMAS EVE-EVE EVER

The response was crazy. Orders poured in. One hundred thousand dollars the first day. The same the next. And the next. Shopify never even broke a sweat under the crushing load of curious customers the *Shark Tank* airing brought to their site. Lots of people recorded the show to watch it days or weeks later, so the "long tail" of the episode's impact made us giddy, a steady flow of consumer interest and real revenue. We'd expected a bump but the bump became the road. By the time the year would end, we had pulled in almost $2 million, lots of that in the six weeks after the show aired. *Shark Tank* lifted us out of our immediate cash crunch.

We knew what we were doing, and we didn't. Two weeks after the airing, the first shipments of DoorBots, 1,000 of them, arrived from Asia. We needed to ship them out fast or risk pissing off the customers who had waited so long since pre-ordering them through Christie Street almost a year earlier. We had built a battery-operated HD camera, something no one had ever done before, enclosed it in a doorbell, and we were getting huge attention, praise, and support for that. There was nothing on the market like it—as I had tried to explain to Mark Cuban during my *Shark Tank* pitch—so even a challenged product was good enough to win, for now. Timing was key. I thought of the legendary warning from Reid Hoffman, co-founder of LinkedIn, to all startups: "If you aren't embarrassed by the first version of your product, you shipped too late."

It was going to be a great Christmas, even as I knew we had a challenged product. DoorBot was almost exclusively direct-to-consumer, through GetDoorBot.com and a trickle of sales through Amazon, and we were about to launch in Staples.

We had moved into our new office, 1523 26th Street, a 5,000-square-foot warehouse/former recording studio with high ceilings, between Broadway and Colorado Avenue in Santa Monica. We shared an interior wall and window with a mezcal company called El Silencio. We were definitely not up to fire code. There were "rooms" that I doubt people were meant to work in. We had set up foldout tables. The air-conditioning never worked. We shared the second bathroom with the mezcal folks. My office was in an attic, reachable by a staircase and catwalk that were so not OSHA-approved. The interior did not bring to mind words like *décor, feng shui,* or *ambience.* More like *asbestos, rats, leaky roof,* and *more rats.* If someone told you that chalk outlines of murder victims had been erased just before we moved in, you wouldn't have blinked.

To call our conference room a conference room was an insult to conference rooms. It was windowless, with an orange-and-tan surfboard tabletop. When I drilled a hole in the wall for an Ethernet connection, sand started pouring out, endlessly, like in a horror movie. The previous tenants, I learned, had filled many of the walls with sand for soundproofing.

We kept a constant supply of Costco-bought peanut butter pretzels, mixed nuts, Red Vines, and Rice Krispies treats. We had a coffee maker but used Folgers. Soon enough someone cracked, "Guys, we can't be drinking Folgers," as if we were actually above it all. August and John upgraded, bringing in a couple of Starbucks Coffee Traveler cartons every morning. When I found out that someone had signed us up for Sparkletts water delivery service, I blew my top and made them cancel it. Did they think we were made of money just because we'd finally started filling orders for our lone product? The roof continued leaking, and one of my team wondered out loud about calling a roofer. I looked at him as if he was from another planet.

"Do you know how to stop a leaky roof?" I asked, sternly. "Trash bags and duct tape." Twenty minutes later, he and I were on the roof, conquering the leak with trash bags and duct tape.

We absolutely needed this bigger space, but I hoped I hadn't messed with garage karma, so I chose to think of our new place as simply a bigger garage. Every morning, when I set foot in my own garage at home, I had felt as if all things were possible, including things I hadn't even thought of yet. It was my cathedral of invention. Some ideas worked out, most of them didn't, but the feeling the Siminoff Brothers had in there was, well, electric. Now, I hoped our new space in Santa Monica was equally good for creativity and fixing things. Of the "Mag 7" tech companies—Alphabet, Amazon, Apple, Meta, Microsoft, Nvidia, and Tesla—four got started in a garage, and a fifth sort of did, too. (Facebook got going in a dorm room.) What does that say about the garage as an incubator of great inventions?

O

As I looked for sources of new money, bigger companies in the home-security field were noodling around, expressing interest in partnering with us. I considered one offer from a well-known lock company that wanted us to build a branded doorbell for them—and then got my sense back and turned them down. Why would we do that? Even in my desperation, I knew that we were a company building for end customers, not manufacturers.

The things you say no to can be more important than the ones you say yes to.

I installed one of the very first DoorBots on the house of my neighbor and good friend Scott Marlette. Within hours, his wife, Stephanie, gave us her blunt review: "It looks terrible."

Scott knew something about technology as one of Facebook's earliest employees, but that expertise wasn't needed for his appraisal. "Jamie, it doesn't work well either," he told me. "Forget the look. Forget the camera. A doorbell has one real function. You push the button and it rings. Half the time yours doesn't ring."

Scott has a kind heart but he's demanding about tech. He wasn't going to give us credit for being a bunch of grunts in a garage, nor should he have; the product had to meet the expectations of the customer and then some, like an Apple product did, or a Samsung product. Like a Dyson product. And $199 was a lot to plunk down for a doorbell, especially one that rings only half the time. It was honest feedback, the type I always wanted to hear, no matter how painful.

"We'll make it better," I promised.

I started hiring people who answered my ads on Craigslist. I put Dave Savage, an eager college dropout, in shipping, at 10 bucks an hour. When he came to me to point out that we were low on boxes, I said, "Great! Order them so we won't be out of them anymore."

"I was hired to pack boxes."

"Not anymore," I said. "In addition to shipping, you now run procurement and planning, too. Congratulations!"

I hired Bill Veilleux, also at 10 bucks an hour, to help pack DoorBots. But I needed to move equipment out of my garage to our new office, so I asked Bill to drive in the U-Haul with me, and we got to talking. He said he was an electrical engineer. He had worked at the Jet Propulsion Lab. He'd done pioneering work with the Steadicam. He wasn't bragging, just stating facts.

"We're paying you ten bucks an hour to pack boxes?" I said.

He shrugged.

"Listen, I'm happy to pay you the same to do engineering instead."

He nodded, as easygoing as could be.

I didn't care what anyone was doing to be answering Craigslist ads. I didn't care if they'd taken unusually long sabbaticals not of their own choice. In fact, I admired it. The fewer impressive credentials, the better, because it meant (a) these people had grit; (b) they didn't have a set way of doing things; and (c) I wouldn't have to pay money I didn't have for the fancy ones. If you were willing to work hard and there was no BS, you had a place with us. I gave John engineering tasks even though he politely reminded me each time that he had majored in industrial design, not engineering—and then undermined his argument by solving the engineering problem anyway. August was my jack-of-all-trades, master of many, mostly taking on business operations, which meant that I trusted him with lots of very important tasks. Whenever the next in a series of CFOs flamed out for us, August subbed in. We were so seat-of-the-pants, his personal bank account somehow got connected to the DoorBot business account, so anyone who had to check the latter could also see how August was spending his hard-earned money. (Turns out he saved almost all of it.) We resembled a pirate ship more than a company.

I contracted some engineers who worked remotely, including a couple of great ones in Argentina. We were a dozen people, then 20, then 30. Every Wednesday, Olga brought lunch to the office for our growing team—at first tacos she'd made at home, along with the best pupusas anyone had ever eaten, then gallons and gallons of chili because it was easier to make in quantity. I wished I were as good at scaling product as she was. She made it look like craft services on a movie set.

Everyone worked their ass off. We could look through the window to the mezcal office late in the workday and see them drinking mezcal and laughing. "Wow, must be nice," we got used to saying to each other. They invited us over for happy hour, and sometimes we would drop by, then return to our office for another four hours of work that probably weren't all that productive. Or we were simply too busy to accept their invitation and were left mumbling to each other, or ourselves, "Wow, must be nice."

○

With more product arriving from Asia and Christmas approaching, it was a mad rush to fill everyone's order on time. We posted pictures on social media showing a DoorBot box poking out of a big envelope covered with USPS postage and captions like, "The first group of DoorBots is SHIPPING!" We were shipping as fast as we could, which was not as fast as our customers wanted. The social media campaign was designed to calm the cancels.

I was asked to speak at a tech gathering in New York City. I felt I had to address head-on what customers were saying about us, so I showed the audience a set of reviews with star ratings equal to those of DoorBot. The most glowing one was still muted: "Good but could be way better." I then revealed to the audience that these were reviews for the iPhone when it launched more than five years earlier. The crowd got it. If the iPhone had earned those types of reviews in the first months after launch, then maybe this hardware shit really was as tough as it looked.

We randomly tested DoorBots before we shipped them to customers—there was no time to test each one—and discovered more problems. Scott's complaint about his doorbell wasn't because he got a lemon or had unreasonably high standards. And it wasn't because he, like everyone else, was comparing the user experience to the one he had with his smartphone, where Apple and Samsung had raised the bar. No, our doorbell was objectively flawed, a euphemism for "kinda sucked." Lots of customers contacted us about connectivity or gave us bad reviews online, though it was often hard to know if it was a problem with the physical DoorBot, the software, or the customer's wifi signal.

While my team worked to fix the bugs, I flew to Taiwan to visit our factory. At the Tatung office in Taipei, I sat down with the team making DoorBots and reviewed the list of issues. We opened one of the devices. I noticed that the antenna, which is usually formed into an inverted "F" for optimum tuning and thus connectivity, was shaped differently. Like... a stovetop. Why was the shape familiar? I couldn't quite pin it down.

"Why is it shaped like that?" I asked.

"That's our logo," said one of them.

I must have misheard. "Excuse me?"

"It's the Tatung logo."

"Wait... you used your corporate logo for the antenna design?"

Someone nodded.

There was no wall close enough to punch.

"How charmed would we have to be that your corporate logo... the physics of it—" —I was practically spitting—"turned out to be *the* optimum shape of a great antenna?" A table might have been overturned. "ARE YOU KIDDING ME?!"

Yes: Someone there had decided, *Hey, wouldn't it be neat if the antenna wires were twisted in a way that looked like the Tatung logo?* Forget that no one but the engineers, or inventors who enjoyed taking things apart, would ever see it. *Sweet, right?*

I hired a group of wireless experts to analyze the entire wifi "stack" and figure out how we could fix the poor connectivity issue. They discovered that the "logo antenna" was only part of the problem; a bigger one was that the entire device was shrouded in aluminum. And who had made that genius decision?

I had.

I loved the look and wanted our customers to "feel" that the DoorBot was a premium product, that cold metal on their fingertips when they removed it from the box. Between that and the connectivity difficulties incurred once you mounted the doorbell on any wall not made of paper, I had basically blocked most of the pathways for a signal to come into or go out of the doorbell. I felt no solace that the design gods at Apple had gone through a similar connectivity issue, "AntennaGate," with the iPhone 4 (when the phone was held in a certain way, you almost completely lost signal). I had gone for form over function and clearly had no idea what I was doing.

DoorBot had a steady flow of technical issues, and they were hardly all the fault of some rogue Antenna Picasso back at the manufacturing plant. We'd been getting reports that the unit would sometimes *ding-dong-ding!* randomly, a problem we were also experiencing in our building. We figured out the cause, but only some of the first batch had that glitch. So if the noise came from somewhere in a stack of boxes, we had to go through each one to find the culprit and "flash" it (update the software) so it would stop.

Other doorbells had a funky chipset that needed rebooting or updating with new software to make it run right. We had to open every single unit, make the fix, put a Magic Marker-ed dot or initial somewhere on the back so we knew it had been checked, carefully close it up, and repack it before it shipped. Not the greatest system to track progress. A shared Google Doc would have been smarter.

Our box read, "Say hello to DoorBot—the world's greatest doorbell!" It was feeling like a hollow slogan.

I got it in my head that if we were opening up DoorBots, we needed a "clean room": a tented space inside our new office with an air-filtration system blowing, where we would open up the unit, flash it, close it, pack it, and ship it out. Like something out of the movie *Contagion*.

It was another rushed solution. I overdid it. To begin with, the devices we were opening weren't really sealed, so there was no reason to be extra-cautious. And it's not like the Siminoff Brothers wore hazmat suits or even gloves when they were inside the clean room. At least Dave (now a member of the Siminoff Brothers too) took his cigarette breaks out in the driveway. As John Modestine put it simply, "I don't think we did a very good job of keeping the clean room clean."

As the holidays approached, we got so behind on shipping, and so worried about customers demanding refunds. More and more emails read something like "if I don't get it by Christmas, cancel my order, [INSERT EXPLETIVE]." For the next shipment of DoorBots (with no more logo-shaped antennas), we didn't bother testing them at all. We just got them out as soon as they came in, hoping *this time* they all worked, while knowing there was zero chance of that.

We had no formal customer support. I was hiring so many people but there were just too many things to do. August set up a Zendesk account and routed tickets to himself from a backlog of 12,000 customers upset that their doorbell had yet to arrive. "We're just updating the software!" he wrote them, being as sunny as one can manage via email. Surprisingly few customers got really angry—yet. Maybe that was because lots of our first purchasers had come to us via Christie Street, so they were likely tech enthusiasts who understood that problems in a version 1.0 are inevitable.

Given the chaos, the complaints, the occasional returns, and the fact that our one and only product didn't work very well, I wasn't surprised to learn that Dave was growing nervous about job security.

"What the fuck—is this a real job?" he complained to John. I'd made Dave Head of Shipping, Procurement, and Planning but technically he was still Head of Boxes and Tape. "I don't think this is an actual job. It doesn't feel like it's going anywhere."

"Give it some time and see what happens," said John, who'd been with me long enough to have some perspective. "Just wait."

Still, the energy in the office every day was raw and passionate, the spirit collaborative. Which was good, because after a very long day and evening, well after our mezcal neighbors had gone home happy, we'd be closing up for the night and shutting all the lights—and all of a sudden, in the distance, would come a *DING! DONG! DING!* from the pile of doorbells we thought we had just fixed.

O

A random *DING! DONG! DING!* was nothing. Or even customers complaining that they hadn't slept because "Yankee Doodle Dandy" or their alma mater fight song or whatever weird ring their original doorbell came with chimed throughout the night with no rhyme or reason. What we didn't realize yet was that all of the units in the next, much larger batch of DoorBots we'd started sending out were broken in a more profound way. Did I say "all"? Yes, all. We just didn't know it yet.

The team was overwhelmed filling back orders, but also now handling an almost equal number of returns from customers whose doorbells didn't work. Something was very wrong. Chris Fralic, the partner at First Round who was leading his firm's investment in DoorBot and who may have been, aside from Erin, my biggest champion despite the troubles I'd put him through, had ordered a dozen doorbells for holiday gifts for his neighbors. He called me to complain that he could no longer walk his dog in peace because everyone on his street was mad at him because their DoorBots didn't work.

"I could understand if one of twelve was buggy," he said, "but all twelve?"

I felt bad for Chris. I hoped that the doorbells he ordered were just a bad batch out of the thousands out there. Who was it who said, "Hope is not a strategy"?

"I'll look into it," I said. What else was there to say?

"Are you going to refund all your customers?"

"I can't."

"Are you going to do a recall?"

I couldn't believe he used the R-word. *Never* use the R-word in retail. "Yeah, I can't do that either."

"What happened to all the money?"

"We spent it all."

The week before Christmas, more and more people contacted me to complain that their DoorBot was not working. I think the technical term is "dead." While our product had experienced multiple technical issues along its rocky development journey, this new problem appeared bigger and more widespread. Sure enough, the video on every one of the most recent batch of units was coming through with lots of lines across it, like on an old TV screen. The picture quality was terrible.

We started testing them. One failed, and another, and another, and another. We'd been shipping thousands of flawed doorbells, every single one.

Most people (which included those I was friends with, related to, married to, or who had invested in me) would say, "Okay, go to Plan B. Now. Take the buggy doorbells out of circulation, take a step back, take a breath, figure out how to do this right."

I understood that side of things. I did.

Problem was, there *was* no Plan B. You need money for a Plan B. No money, no options. For us, the only way was *through* the problem. Plan A was fix it. Plan C was, you're dead.

We knew the scale of the problem; now we had to figure out the root. And at the same time, sell the remaining DoorBot inventory and get the money from that to build a better doorbell and camera. I couldn't slow down to take a breath.

The day of the eve of Christmas Eve, we finally figured out what had happened.

Weeks before, a programmer I'd hired (through a professional recruiter, *not* Craigslist) said he had a way to make our decent video quality way better. We were all behind the effort; we all knew it needed to improve. The programmer changed a few settings in the code, tested his hypothesis, reported his great results, and sent a new software build to the factory in Asia. There, they embedded the new settings onto the camera chip and shipped those DoorBots to us. And because we were so behind, we didn't test them before packing the units into boxes and shipping them out to customers.

Now, I sat with the programmer who'd thought he had *improved* the camera's video quality, not made it far worse, and we realized that when he was testing his software at our 1523 office, it wasn't actually "writing" to the unit. He'd made a mistake, one big enough that it was about to end the company. His blunder: He didn't comprehend that when he got the picture to look really good, it was because the internet connection he was working with was particularly strong; conversely, when the picture looked bad, it was because of a weak connection. He took the settings when the internet was strong, reprogrammed the software with them, and sent that off. In short, the camera settings he sent to the factory were based on nothing: basically, a random set of numbers. And those random settings got "printed" in the camera chip's firmware.

Not only did it mess up the video, but another mistake compounded things. It turned out that the chip was incorrectly wired, and it was the only one on the DoorBot that could not be reprogrammed through an "over the air" update (an updated file you send to the customer after install). What was on the chip from the factory was burned into the chip. Those factory settings were never changing. The problem seemed unfixable. Not by software. Not by taking the device apart in our fake clean-room tent. In industry language, each DoorBot was now essentially a useless "brick," a word in the hardware startup world almost as terrifying as "recall."

It was the eve of Christmas Eve, and we had shipped 5,000 doorbells with bad, unfixable cameras to eager customers. More horrible reviews, which we totally deserved, were starting to appear.

be good if I could make out who it was – my mother or jeffrey dahmer

Device took too long to open a video connection on my phone.

Goes through batteries too quickly.

DoorBot app lame not great user experience

TOTALLY FUCKED UP VIDEO. WHAT IS THIS, 1950??

I couldn't be angry if one of Chris's neighbors had written one of those reviews. Or if Chris had.

I had no money to issue refunds.

Almost $1 million worth of broken product had just been delivered to our customers. Those units were toast. There was no solution any of us could see.

In that horrible moment, I got eerily quiet. Methodical. I went to the checklist. "Pilot mode."

Airplane pilots are incredible in many ways. One, they're flying a piece of metal filled with people through the air at over 500 miles per hour and, as a group, their passengers are safer than pedestrians walking on a sidewalk. When something unusual happens in flight (or even when they're simply confirming the usual), pilots go through a checklist, and the most effective way to do that is calmly. The wing just ripped off and the aircraft is falling out of the sky? Go to the checklist.

I spent the eve of Christmas Eve going through the checklist with the team. We tried everything we could think of. Could we heat the chip, pull it off the board, reprogram it, then lay it back down? Nope. Could we "jump" it to get the correct code on it? (Just like jumping a car—hook wires to it and a burst of voltage *might* clear the chip.) Nope. We went through every sane idea, then every wacky one. The answer to each was the same. I tried to maintain pilot mode but I could start to make out the outline of the mountain up ahead. It was now close enough that I was pretty sure we were going to hit it.

That night, Erin, Ollie, and I went to dinner at one of our favorite places, Houston's in Santa Monica. I remained calm, still in pilot mode.

We had not officially hit the mountain. I kept reviewing in my head the checklist. Could I have missed something? What about this... or this... or this?

No, no, no. There was no way out that I could see. Neither could Mark Dillon or the other engineers.

The only conclusion I kept coming to was the most obvious one: *We're done. We're going to hit the mountain.*

It made me even more eerily quiet. Erin was watching me like I was a chemistry experiment that was about to explode but, for inexplicable reasons, didn't. Yelling at the clouds or lamenting bad hiring decisions wasn't going to help. Or punching walls. I didn't think Ollie could tell something was up. No, probably he could.

At the tables in the restaurant all around us, families and couples were in good moods. The waitstaff seemed in a good mood. Why not? December 23rd may be even more fun than Christmas Eve and Christmas itself, because it's 100% anticipation, not yet the actual execution of the holiday. All the good things were still to come. Nothing bad had happened yet.

It had been a fun run. Sad for it to be over, but fun. I told Erin it was over. She didn't believe me. I told her that there were two options, both terrible, both financially unworkable: one, take back thousands of doorbells and refund money; or two, stop production, eat the cost of the bricks, and start a new run of (hopefully) working units that wouldn't get to people's homes for months, way after the holidays. In both cases, I was dealing with floating way too much money, having nothing coming in, getting skewered for our terrible product, and in the meantime watching copycats or better-funded players swoop in and likely learn from our self-inflicted stupidity. Also, no one was giving us money after this debacle.

I told Erin it really was over—*really*, no kidding—really over before it had started. It felt like getting injured 30 seconds into the start of Game 1 of the playoffs. It was the first time you'd made it there and now it looked as if it would be the last time, too.

"It's not done," she said.

"No, it is, it really is," I said. I even smiled. I didn't want to ruin our dinner, too. "It's done."

"Why don't we mortgage the house?"

I looked down at my lap. I felt tears coming. I shook my head. A dream I'd had before came to me: I'm in a boat heading into the worst storm ever. I'm going to sink, no question. There is no hope. And then I realize there's someone next to me in the boat—Erin, even though I don't want her anywhere near this boat in this weather—and she says, "Let's go *into* the storm."

I shook my head again. Tears rolled down my cheeks. It was a beautiful suggestion but ultimately unworkable: Our mortgage would not come close to covering the hole we were in.

I thought I had given up on finding a solution, but I still had that one not-so-secret weapon. No, not an Ivy League degree. The fact that *I never stop*. Until my last breath I'd be thinking about it, even though I was certain this DoorBot adventure of mine was done.

I tried to enjoy the food, a zen exercise. Ollie made it easier. Erin, too, though I could see flashes of fear and sadness in her eyes. She had put on such a brave face. I kept smiling but couldn't keep another scene from playing in my head.

The screenplay version:

WE'RE IN A BAR. IT'S NIGHT. IT'S FIVE YEARS LATER.

The bartender, JAMIE, 41, is wiping away spills up and down the bar, deferentially working around a couple of CUSTOMERS.

> CUSTOMER #1 (to friend)
> You know this video doorbell thing?
> It's amazing.

Jamie pauses. He wants to say something, then stops.

> CUSTOMER #2
> Oh, I know. I've got one. I love it.
> Incredible.

It's too much for him.

> JAMIE
> Actually... I invented it.

> CUSTOMER #1
> Excuse me?

> JAMIE
> Yeah, that was my invention.
> The video doorbell.

The customers look him up and down, like he's in a police lineup. Their eyes say, *How sad.*

> CUSTOMER #2
> No way. Billionaire Joe Shmo invented it.

> JAMIE
> Nope. I did. The first one. I did it because
> it made people feel safer. I had a company
> and everything. Joe built his after I flew
> the company into a mountain Christmas
> Day 2013. But I'm not bitter. His company
> actually built the working version. Good
> for Joe. I'd love to meet him someday.

Nope. The customers are totally not buying it.

JAMIE
Look it up.... How about a round of shots
on the house, fellas? Mind if I join you
for this one?

On the drive home, Ollie sleeping, I kept chewing on the problem. And chewing. Even in my greatest despair, I couldn't stop.

A thought came to me. For some insane reason, it had evaded my pilot-mode checklist and bubbled up only in that moment:

Weeks earlier, the system that processed the video in the cloud before it went to the DoorBot app had changed. We had swapped out the server software (because: no money) for a cheaper one (which did only the one thing we needed, not all the other things the expensive server did). Could it be, I wondered, that the *previous* server could actually process whatever video was being produced by the faulty units we'd shipped, and clean it up?

I called Mark Dillon in New York. It was very late there but not the first time I'd woken him, or the twentieth. I shared my thought with him. Well? Did he think the more expensive cloud software we'd ditched could read whatever was coming out of those cameras? And correct it?

Trying anything was better than trying nothing. He was on it. He said it would take him at least eight hours to get it up and running and tested. I could have apologized for waking him up and making him work all night but it would have been insincere.

My phone rang 6:30 Christmas Eve morning.

"Dude," said Mark, exhausted and wired. "It works. It fucking works!"

Mark never swore. Our Hail Mary had actually fucking worked! When we had updated the video-streaming software package on the new server, it had lacked the adaptability of the more expensive software, and could not automatically adjust to the problem the engineer had caused by changing the camera settings. But the previous package could. With the better server, Mark was now looking at the best video quality he'd ever seen. "Merry Christmas," he said.

It was the best Christmas the Siminoff family ever had. An hour later I sent an e-blast to every customer. Forgive my generous interpretation of the word "few":

> A few of our customers may have an issue with the picture quality. We have made some updates and you should now be fine. Happy Holidays!

O

We were in better shape than before, for sure. Orders coming in, money coming in, an existential threat sideswiped. True, with all the broken doorbells we'd shipped, the reviews online were pretty terrible. But our spirits were high, the DoorBots were starting to actually work (ish), we had stared into the face of death and were still standing.

But we had to up our game—not just our product but us. Everything we did. For one, we could never again put ourselves in a situation where we could not easily update the chip. We had used hacks and shortcuts and seat-of-the-pants solutions the first time around, but we wouldn't last long if we continued to operate like a bunch of amateurs. It was one thing to feel like a ragtag bunch doing its best in our ugly, overcrowded, non-air-conditioned building. In the end, the outside world doesn't care about your cute founder story, nor should they.

We all committed to being the best in everything we did: product quality, design, benefit gained, patent-worthy innovation, on-time delivery, ease of installation, affordability, customer support.

Because if we *didn't* do that, then we couldn't fulfill our mission, which was to make people feel safer. To make their homes safer.

If the word "mission" sounds grandiose, here's the thing: I don't have moon-shot ideas. I really don't. My friend Diego, who has more of a gambler's mentality, does. I have lots and lots of little ideas, tons of them. Most of my product ideas for Edison Jr. were things you could imagine seeing on QVC or in the SkyMall catalog.

But little ideas are sparks, and sparks become embers, and embers start fires. And my little wish to not miss someone at the front door when I worked in my garage had turned into something bigger. Erin immediately understood what a video doorbell could do and the impact it could have. I was determined to build a product that lived up to her aspirations for it.

A rebuild of DoorBot was what we needed. In the week between Christmas and New Year's, on the cusp of 2014, I gathered the team and laid out a plan, codenamed "F5," which stood for "fucking 5-star reviews." We needed to go harder, smarter. We should aspire to be as elegant and satisfying as products from companies whose quality, innovation, reliability, and especially obsession with customer satisfaction we admired, like Apple, Samsung, Dyson. "If we don't rebuild everything, the back end, the product, the camera, the packaging, the name," I told them, "if we don't show what this thing can really do, and don't launch it by next year's holidays, we're done. We deserve to be done. But I know we can do it."

Five-star reviews were not about just our product, but the perception of our product compared to everything else the user experienced. When you pay $200 for a device at Best Buy or Amazon.com, you compare it to other $200 devices, to similar things you own. In our case, that was likely the iPhone. At first, we'd gotten some grace from a handful of early buyers when they found out we built this thing in a garage, but we were way past that now. My team understood that we were now competing not with another upstart, not some garage copycat, but Apple. The rats in the bottom of the ship could eat each other. We were going up top, where the battle for supremacy would be waged.

I wrote up a manifesto that began:

DoorBot V2
Extremely Confidential Shit

Version 2 of DoorBot will need to include many of the same features core to version 1 while enhancing the unit and shrinking it in size.

Current necessary minimums:

- All chips need to go from deep sleep (off) to fully awake and loaded in less than 1 second
- IP54 or ideally IP55 (waterproof)
- Dual power (battery and 8-24 volts ac)
- Temperature range 0-130 Fahrenheit (at a minimum)
- Night vision

I went into detail about every key issue: electrical, mechanical, cloud, the bell sound, patents, more. Everything needed to be worthy of five stars, I told them. *Everything.* I was particularly concerned about improving and innovating on motion detection. DoorBot didn't have motion detection, and Erin said she "felt safer" because she could see who was coming up to the house. It did half the job. It captured the prospective burglar if they pressed the doorbell, which from our anecdotal data was about 50% of the time. Motion detection would cover the rest of it, for all those creepy characters who *don't* ring. We promised to make people safer. That was our mission. "This fix is not about the doorbell," I told the team. "It's about who we are."

If we could deliver F5 by October 1, we could own the 2014 holidays and entrench ourselves before a serious, well-funded competitor popped their head up. Miss that, and I was pretty sure my future was in bartending.

If DoorBot was the alpha product, then F5 was skipping right past beta and going straight to platinum record.

○

In the last days of 2013, we learned about the technical challenges still before us but also about customer expectations, which would help us with future design and business decisions. For all the early adopters and tech enthusiasts who gave us the occasional mulligan, August on his Zendesk account was getting way too many versions of "Yo, you have to get this product to work or my wife is going to make me take it down" and "My spouse *hates* the way the doorbell looks."

So... what *should* our doorbell look like? Did it *go* with most housefronts? How could we make 100% sure it correctly performed all the basic functions it promised? Were we wasting money on features people didn't need? What was the essence of this new doorbell? While I was hyperfocused on the mission, we could only achieve what we wanted if lots of people bought and installed one.

Being on *Shark Tank* had provided awareness and an influx of quick cash, but it hadn't paved the way to a big investment as I'd hoped. Sales were tapering. Word was spreading that DoorBots were not perfect. Many in the investor community at the time looked down on *Shark Tank* and the entrepreneurs on it; to them, we were reality TV stars, not hardworking inventors and businesspeople. With our record of real sales, though, I thought it would be—should be—easier to raise the capital that every startup needs.

Nope. To many of them, especially the ones that had graduated from Ivy League schools, I was just a silly doorbell salesman.

CHAPTER 4

YES, MARK CUBAN, BURGLARS RING DOORBELLS

Back when Diego and I were thinking about smart home-security systems, I'd spent time researching the psychology and habits of the residential burglar, the cat-and-mouse game of Inventor figuring out what Thief will do and how to make that job harder. I studied burglar mentality, which was at once fascinating and also totally obvious.

For one, burglars ring doorbells, as I pointed out to Mark Cuban on *Shark Tank* when he thought he had me in a *gotcha!* moment. Burglars *want* to see if you're home. They want the opportunity to be in your house *when you're not in it*. Burglars are not usually violent criminals. They almost always work during the day, when you're way more likely to be out. And they're not just looking for an empty house but an empty area. They will ring multiple doorbells, but three are crucial: the house to the left of the one they want to break into, the house to the right, then the target in the middle. Very Goldilocks and the Three Bears. Makes total sense, right? Burglars aren't stupid all the time. They don't want the neighbors hearing or seeing something and calling it in. If even one of those three homes answers the door, the burglar is probably out of there. Maybe they'll try another street or neighborhood, but not yours. There's a real positive to creating a presence in your home whether you're there or not.

And it's not as if the pursuit to reduce crime and make the world safer ever ends. An inventor never thinks the job is done. Crime to an inventor is like catnip, a forever problem. The thief always adjusts, so the inventor needs to as well. That's what makes crime-solving stimulating. They adjust to your adjustment. You adjust to theirs. You want to come up with something that makes you safe, makes your loved ones safe, your home, your street, your community. Maybe you come up with a system good enough that the thief decides, *Fuck it, maybe I'll try working for a living.*

As we headed to the Consumer Electronics Show (CES) in Las Vegas in January 2014, I felt more frustrated than I ever had as an inventor, an entrepreneur, anything. We had already kicked off F5. We knew what we wanted to get into the market nine months from then. We were so psyched about how good it could be. But instead of talking about that at CES, we had to smile and pitch the greatness of DoorBot.

It's one of the hardest adjustments for anyone in the product business. You know what the new, better future is, but because it takes a year or more to build and launch, you're always pitching the current, antiquated product as the best thing out there.

In this case, it actually *was* the best thing out there. Yes, DoorBot was flawed, but it was still a pretty incredible feat of engineering compared to anything else on the market. So while we worked on the next, 10x better version of the video doorbell, we were probably too critical of ourselves. And maybe that was the secret to success: If we spent energy congratulating ourselves on a product well done, it would take away from the energy needed to make the next product so much better.

Driving in my car to Vegas with Luciano, one of my engineers based in Argentina, I thought out loud about what my research had told me: how our doorbell could protect not just you but also your neighbors. And if your *neighbor* got our doorbell, then they'd be doing the same—protecting themselves and you. I began to play with the concept of a "network of neighborhoods" and a "neighbors app," which would connect everyone to some hive mind for greater safety, and in the process create a narrative around our little gizmo. That idea worked only if we hit a tipping point of acceptance. If we *did* reach that, though, there would be a multiplier effect, where people started to believe that one new smart doorbell out there named DoorBot was simply better than all the competitor doorbells, true or not. With enough market penetration, our doorbell could be that. Luciano and I discussed a video-request tool—where after a crime, the police could send out an email query to see if anyone had video to help bust the criminals, in a way that also protected the homeowner's privacy.

The more we pulled at the thread of the idea, the more it seemed like something we could work on forever. We delivered *presence* to homes that didn't always have them, at least not for just $199. We were not *selling* doorbells. Our customers—"neighbors," as I thought we should call them from now on—*rewarded* us with the opportunity to provide them with a sense of safety. I needed to plaster the mission onto every team member's brain, every corner of the office. *We exist to reduce crime in neighborhoods.* That's it. That's what would make us or break us. Letting the mission guide us. The mission, nothing else.

It was my epiphany. My aha! My breakthrough. My eureka! Call it anything you want.

It was a really big deal.

I should have understood it the instant Erin first said what she said about the doorbell, but I was too busy solving my small, practical problem. Now, I realized we were actually solving a huge problem, practical and psychological. Local and universal. For now and forever.

That was it!

Luciano and I were psyched to go deep on the whole concept—but it was yet another source of frustration, because our team could not share any of it at CES. No way. Not F5. Not the neighbors app. And certainly not the whole mission behind what we were doing. It would all have to wait. Not because I was worried about copycats or idea thieves, but because we had no credibility. You have to earn some trust before you can say you're going to change the world and not be laughed at.

Luciano and I, and the rest of the DoorBot gang in the caravan heading to Vegas, hardly felt special when we got to the venue. We'd signed up late for CES and were given a "pity party" booth, free but way smaller than the standard size and located in the least-trafficked area of the Venetian— the ballroom upstairs, where no one went. Not that we could afford anything bigger or situated better. Someone called it "Dracula space" because it was as far from sunlight in the middle of a desert as you could get.

It wasn't the first time I'd been in the boonies. I'd been coming to CES on and off since 2006 and my Simulscribe days. For our display then, I'd made two giant printouts, one each of a Palm Pilot and a Blackberry (this was pre-iPhone), a blank screen on each. A cheap projector behind each screen played an animation of how Simulscribe worked. Most of the people who stopped by our booth were more interested in the display I'd invented or knowing how we'd gotten those personal digital assistant giants to sponsor our display (we hadn't) than in what our product did. I'd made an $800 booth look like a $20,000 booth.

Now, eight years later, I was back in Vegas, again in the conference boonies, needing a way to get attention. We stacked empty DoorBot cartons behind us to give the appearance that our booth had actual boundaries. Everyone around us was going futuristic and high-tech with their booths, so I thought we should do the opposite: make it cozy and conservative, even retro. We bought some green turf for a nice, homey feel. We brought in a small live tree.

I hoped that would get us some attention. Our section was for the afterthoughts and losers. Most of the people at the booths looked the part, slumped in chairs, half-asleep. I couldn't blame them, but I banned chairs from our booth. We all stood, always one of us out in the aisle, forcing people to see us as we "invited" them to come see our booth, where I gave them the world's quickest elevator pitch, talking even faster than usual.

As we walked through some of the brighter, more spacious venues and high-traffic spots most presenters enjoyed, we looked at each other and said, "Wow, must be nice." If it was not clear before CES, we realized now that it was existentially imperative that we hit the October 1 date, when F5 would launch and soon after become the market leader. Looking around, you could feel the shift: CES 2015 was going to be all smart thermostats and smart smoke detectors and smart doorbells. We had to own the front door before the rest of the market could catch up.

The good news? We knew the launch window we needed to hit to be successful.

The bad news? Any sane person told us it would be impossible to hit it.

At the biggest electronics show in the world, I wanted to tell everyone about the amazing things we were working on for the future, but that would have to wait. Instead, we pitched the hell out of DoorBot, the flawed product I'd mentally moved on from. We impressed enough people to get a ton of media coverage, got highlighted in BBC News's coverage of the show, and won *Wired*'s "Best of CES 2014."

As soon as we got back to LA, I drove to Kinko's (remember them?), printed out a set of large vinyl stickers in blue that read "TO REDUCE CRIME IN NEIGHBORHOODS," headed to the office, and planted the stickers on the wall of the warehouse, so they were the first thing you saw when you entered.

A week later, Google bought Nest Labs for $3.2 billion.

CHAPTER 5

THE TREADMILL

Eight million bucks.

That's what I calculated it would take to do version F5 right. Our mission, all the ideas about safety and crime prevention and neighborhoods that I'd been refining with Diego and then on the trip to CES with Luciano, would certainly make for a stronger fundraising pitch.

As we brought on new team members, I always shared our mission. On some faces, I could see eyes glaze over. I got it. Every big company has a "mission statement," and it's almost always some epic goal expressed in the blandest language, doesn't matter the industry:

We make Earth a better place.

We do good things for nice people.

Our stuff makes people happy.

But when I told them our mission was "to reduce crime in neighborhoods," they seemed to light up. The aim was both ambitious and specific, a goal you'd be proud to share with your out-of-town friend over a beer when asked to describe what exactly it was that your crazy startup in the rundown building was trying to do.

And for those whose eyes were still glazed over after hearing my talk: They were not for us. "I appreciate if you don't believe we can do what I'm saying," I told them. "Hell, you might be right. But I only want believers here."

I sharpened the mission further with my entrepreneur friends Saar Gur and Matt Brezina, and the tech writer Om Malik. Matt, who'd co-founded an apps and services company called Xobni ("inbox" backward) and who had also invested in Edison Jr., talked me up to a Silicon Valley venture capital firm called True Ventures. They intrigued me because, unlike so many firms, they didn't shy away from startups that did hardware, and had an impressive record with them. They were the first investors in Fitbit and Peloton, and had supported close to two dozen other hardware concepts.

I was introduced by phone to Adam D'Augelli, the newest member of the firm. He and I hit it off—it was immediately clear he wasn't like so many of the other VCs I'd met and pitched before—and after I explained DoorBot's ultimate vision of reducing crime in communities, and that we'd already taken in decent revenue, he seemed genuinely psyched about being in the doorbell business and the impact we could have on the world.

Then it was time to talk numbers. "I'm raising six to eight million at fifteen million, post-money," I told him, a number I'd mostly just made up.[1] It was a lot of equity to give away so early, but when you're more or less insolvent, any deal is better than none. The *Shark Tank* offer from Mr. Wonderful seemed far, far in the rearview.

"We don't do eight," said Adam. "For an A round, True Ventures typically does between one and three million, and even three is on the high side."

"Fine," I told him. "I'll take three million."

Now, if *I* were Adam, I would have considered that a flag about as red as it gets: *Entrepreneur, without pausing, agrees to less than half of what he just said he needed.* Adam didn't know that we were close to broke, or the number of VCs who'd already said no to me. I myself had lost count. I hoped he wasn't pals with Ashton and Mila.

---

1   "Post-money" refers to the company's value after the money raise for that round. If the company was worth $3 million and you raised $2 million, it's valued at $5 million, post-money.

Adam said he would be driving through Joshua Tree the following weekend, so I invited him to LA to check out our operation. He came by 1523 26th Street and we talked non-stop for two and a half hours, all about ways to reduce crime. I told him we wanted to be a device that stopped crime *before it even happened.* Pre-crime. I told him about the messages we were getting from customers—"neighbors," as we now called them—who'd had suspicious visitors and wanted a copy of the recording, for themselves or the police. That need could mean a very significant recurring revenue opportunity, as a monthly subscription fee to store users' DoorBot video in the cloud. I knew, as Adam's firm surely did, that since hardware is often a one-time purchase, we needed other revenue streams, at least until our company expanded and diversified its offerings. And a monthly subscription was not just good business: It bonded customer and company beyond the purchase. It signaled a commitment, a vow to keep up the service. And neighbor satisfaction was everything.

To persuade Adam more, I reported that we had just gotten very lucky with Arrow Electronics, a Fortune 100 chip distributor. One of their reps, Billie Madha, so believed in what we were doing that Arrow had given us a large credit line, plus generous payment terms on the components we bought through them (120 days). It always helps if you can tell someone who's *thinking* of believing in you that someone else already does.

I sent Adam, looking half his 26 years, back north with a doorbell to convince Jon Callaghan, his firm's senior partner and co-founder, that it was worth meeting me. Adam was new to venture capital, while Jon had earned respect after decades in the business. As soon as Adam left the office (he told me later), he called his boss. "Jon, I just got out of a meeting at DoorBot," said Adam, "and we have to invest."

"Doorbells? I don't think so."

"I'm bringing a product back. You'll put it on your house and see."

Back in the Bay Area, Adam presented the DoorBot to Jon, who, along with his son, installed it on his house. The duo worked hard to get it to work but, like many of our neighbors, they ran into trouble.

When Jon's wife came home and saw the doorbell, she said, "That's the ugliest thing I've ever seen. Get it off my house."

Adam called me, panicked. "Jamie, I got you on the schedule for our partners meeting next Monday." VCs often do a weekly meeting with all the partners: the first half to meet with prospective investments, the second half a closed-door discussion to decide which to invest in. Of course I said yes to going up there, but I heard later that Jon saw my name on their visitor calendar and angrily phoned Adam.

"Call him back and tell him not to come," Jon told his protégé. "There is absolutely no way we're investing in this company."

Still new to venture capital, young Adam stood his ground. He explained to Jon that other firms simply didn't believe the physics and engineering of what we were doing was possible. Part of their inability had to do with a fixation on cameras that were hardwired, not battery-powered. So many investors passed on us immediately because they didn't believe we could produce consistent, quality video with battery-powered wifi. I couldn't blame them. As a group we didn't exactly look like NASA material.

I knew from Adam that Jon was comfortable investing in early startups and unpolished entrepreneurs, but we pushed the boundaries on that, given that his single experience with our product was less than impressive, and that it had blemished the front of his house. But we were pursuing something way bigger than a doorbell. "Jamie is sort of crazy, yes," Adam acknowledged to Jon, to make our case stronger, "and now he thinks he can do this on three million from us, not six to eight. It's worth at least meeting him."

Give Jon credit. He begrudgingly agreed to the meeting, in True's beautiful Palo Alto office. I hadn't even sat down when he said, "So my son and I tried to install the DoorBot on my house this weekend."

"I'm guessing it didn't work great."

"That's an understatement. Your doorbell basically broke my house. It doesn't work consistently. My wife thinks it's ugly."

The only good strategy was to double down. Triple, quadruple. Which wasn't hard, because my own belief was growing. Before he could pile on with more criticism, I said, "Jon, you're absolutely right, but let me tell you what we've learned about the product and the technology and why we're excited..." For the next 20 minutes, I pitched the bigger story of what our doorbell could do for him, his neighborhood, his kids' neighborhoods. We had a mission: reduce crime in neighborhoods. I said that his money would allow us finally to build the quality product we'd been dreaming of, one that could live up to our bold mission.

As I spun my tale, I realized how much easier it was explaining to money people the bigger play with DoorBot than the stories I'd once told around PhoneTag or Unsubscribe or other ideas I'd dabbled in. Consumers had "gotten" those concepts, but was there ever a big enough market for them? (Apparently not one that I could find.) VCs are always looking for whales, unicorns, elephants, pick your beast. And now I could spin a tale, a true tale, that *we were selling safety*. Convenience, too, but every tech product or service sells convenience of one sort or another. This was something more. I said I wanted a DoorBot on every door, and since 50% of doorbells in America can't be connected to the home's electrical system, only a battery-powered product could help make these neighborhoods safer. "And," I concluded, in grandiose style for a company that so far had only made mediocre doorbells, "we alone do that."

Adam, the youngest and newest member of the investment team, followed with a passionate defense of how we'd all gotten there. This was now less about trying to convince Jon of the rightness of the idea but finding an opportunity that Adam had specifically been trained to find: flawed, on the cusp, big upside, mind-blowingly exciting. The possibility of doing good made it even better. He talked about DoorBot having the huge advantage of "unfair mindshare" in a developing market. Adam understood that our focus on the doorbell was not the endgame: We were doing security and crime prevention, which could lead to multiple products. To network effects. To some sort of software subscription business. Adam and I saw the world unfolding the same way; I was glad he was taking the reins on what he and I had discussed in LA. I repeated to Jon that if True gave me the money, I would build a wifi video doorbell with everything in it that I had promised.

"You're something, Jamie," he said, "but no, you can't do that. And launch on October 1. It's not possible."

He was 100% right. It wasn't possible, October 1 was less than 10 months away. Actually, it *was* possible but only if you were willing to break every rule, break yourself, and break some of the people who worked for you. "Jon, give me the money and I will launch exactly when I said. First of October. I will not let you down."

I was calling Jon's bluff, same as he was calling mine. I could see he loved our mission. He seemed to like me. But he also wasn't going to allow me to come into his office, sit across the table from him, bullshit him, and expect to be taken seriously.

He looked at me for a long time. I respected that his objection seemed purely pragmatic, not a lack of vision, like Mark Cuban's shortsighted dismissal of the wifi-doorbell market size on *Shark Tank*.

"Okay, you got the money," said Jon. "October 1."

Unlike most VCs, True didn't need the closed-door meeting afterward. They shared our bias for action and went all in.

Jon was backing the mission but he was also saying, more or less: *I'm giving you the money just to show you you're wrong.* He didn't want me to be wrong, because he was a decent person and a fiduciary to his firm. But even decent people can't look away from a disaster. The phrase "rubbernecking" wouldn't exist without lots of good people slowing down to check out the carnage from an accident.

Jon won either way. He had to see the cards. No way was he folding.

Outside, Adam and I hugged. I owed him because from the moment we'd first talked, he believed in me and the team, the product, the mission. From the start he had asked great questions. He was incredibly smart, he always wore black, and frankly he was a bit weird, for which I was extremely grateful. Why? Because that weirdness was going to make him successful. VCs are on the lookout for gaps and opportunities and unmet needs that no one else sees. If everyone could see them, there'd be no need for venture capitalists. Conventional-thinking non-weirdos make bad VCs, generally, and do well only when the tides rise so fast that *everyone* makes it, despite themselves.

But to succeed long-term in venture capital? Bring on the weirdos, please. At 26, Adam had a form of x-ray vision—and I'm not saying that just because he had the foresight to believe in us back then. Adam and I would speak nearly every day for the next four years.

O

Three million bucks was great, don't get me wrong. But it wasn't going to get us to where I had promised we could get. To do it right, we needed another $3 to $5 million. Diego and Sky Dayton, a friend and venture capitalist, made an introduction for me to three very smart VCs who had invested in multiple sensor companies—Chamath Palihapitiya, Mamoon Hamid, and Ted Maidenberg, co-founders of the firm Social Capital. With their interest in the smart home, they seemed perfectly positioned to invest in us. And unlike True, they often invested bigger chunks, more like the $7 to $10 million neighborhood. They said they could meet on Saturday morning, also in Palo Alto. I flew up there, we had a great meeting... but Chamath emailed just days later to say, with regret, that they weren't ready; they had to see how things unfolded with their own companies. I emailed thanks, expressed my own regret (I was seriously bummed), and included a link to a story about how the police up north in Los Gatos were asking residents to register their surveillance video systems. If that was the future (and privacy concerns would need to be considered, of course), then whoever owned those cameras would do very well and be a key player in reducing crime. I sent the story to Chamath because it was relevant for a sensor company, but if I'm being honest, I was also leaving a paper trail wherever I went, for all those who passed on us and would one day regret that they'd missed the boat.

But every down was countered by an up. (Unfortunately, the reverse was true, too.) Around this time the *Wall Street Journal* interviewed me for a story about entrepreneurialism.

> **WSJ**: What's the hardest part about building a startup?
>
> **Mr. Siminoff**: That it takes seven years to really build. Some are lottery tickets and go a lot faster, but to really build a company takes five to seven years. You have to be able to focus and take bite-size chews. If you get too hyped up and [focus too much on] what the competitor is building, you just end up burning out of your cash and going out of business. Being able to focus and execute is the most important thing.

I hired some great engineers, like Tim Simons to lead hardware development and Trevor Phillips to lead embedded software. They had been government contractors and told me they were qualified but couldn't tell me what they had actually just done. I later learned that they had devised a system to detect roadside bombs (IEDs) in Afghanistan, or anywhere else, saving the lives of countless American troops and others. Making video doorbells was unlikely to overwhelm them.

Soon after they started, I introduced them to Bill Veilleux, the former electrical engineer who had found his way to me via Craigslist and was only too willing to drive a U-Haul. He was still working with Dave in shipping.

When Tim got me alone for a moment, he said, "Are you serious?"

"Dude, he said he's an electrical engineer," I said.

Two days later, Tim poked his head in my office. "Bill is one of the best engineer techs I've ever worked with." I gave Bill a raise and made him full-time. Thank you, Craigslist.

I assigned one of our team to buy every single type of doorbell chime he could find—at retailers, online, hardware mom-and-pops, anyplace. We wanted to design the circuit for our doorbell to be compliant with anything out there. We set up a wall of doorbells that looked like an art installment. I don't care how much you may love being in the doorbell business, hearing them go off over and over gets annoying. But it was nothing you couldn't fix with a little duct tape (my go-to for half of all fixes in the world): You could hear the chimes trigger but the sound was muted, and office peace was restored.

I agreed to speak at a conference in San Francisco sponsored by Orange, the French mobile-phone company. At this point, DoorBot was still a speck on the landscape of smart homes and the "Internet of Things" or "IoT." (IoT was the favorite term for the world of the near future, where all sorts of devices with sensors—thermostats, doorbells, fitness trackers, etc.—would connect and communicate with one another.) Since hardware was getting hot, the conference organizers found me. In the waiting room before my presentation, I got to talking with another of the featured speakers, Mark Suster, a VC from a firm called Upfront Ventures. We played the name game; he knew my friend Josh Roth well. When Mark realized I was an entrepreneur, he asked about my latest passion, and I told him.

"What's a DoorBot?" he asked.

I explained.

"Why does the world need a video doorbell?" he asked.

"To feel safer," I said, as if it should be apparent to everyone. Even if this wasn't a formal meeting, I was in the presence of someone with the power to deposit money into my business account, so I recounted our origin story and talked about how we wanted to create a ring of security around neighborhoods, and all the benefits that would come to individuals, families, communities. Not just crime prevention but a tighter, more cohesive society. I told him how we had to be thoughtful on privacy issues. I joked that the camera was *not* for catching the man of the house and the nanny making out on the front steps. I brought up the subscription service to save video to the cloud for a monthly fee, though it was still just an idea. I told him we'd recently gotten a sizable commitment from True and were looking for more.

"Are you a phenomenal CEO?" he asked.

I smiled, then shrugged. Was it a trick question?

"Because nothing that True does, nothing that any other investor does, is going to determine your success," said Mark. "That all comes from the belief and commitment of the founder."

"Agreed. And yes. I am."

"Well… that's it," said Mark, reaching to shake my hand. "I'm in."

Upfront was in for a half-million dollars, maybe more down the road.

Several weeks later, I visited their office in Century City to give Mark, his partners, and other team members a DoorBot progress report. I'd been warned that one of the attendees "could be difficult." I confidently pitched to the group, including Mark, who had made it seem as if investment on top of the initial half-million was more or less a done deal. As I finished, one of his younger team members asked, "How does the doorbell work in the cold?"

"It works fine," I said.

"How about, like... forty below?"

"I don't know," I said. "I haven't tested that yet."

"What about twenty-five below?"

VCs like Adam and Mark ask great, useful, penetrating questions. Some VCs ask idiotic questions.

"Not that either, yet," I said.

"Well, there goes Chicago," he said.

My interrogator started name-checking North American cities that get really cold, ergo the entire wifi-camera doorbell enterprise had a big problem. After listening to shoutouts to Minneapolis, Green Bay, Denver, and Vancouver, I said, "I get it," though I wanted to drop an f-bomb or two. Of all the things we were doing wrong, what a place to focus. He could have questioned me about any of a dozen legitimate, foundational problems with our product. His fixation was an issue that just did not matter. I told him that we had tested the doorbell in extreme conditions but, granted, not quite as extreme as he needed. I would do that, I promised, and find a way to resolve the cold-weather issue, if it even was an issue, and thank you so much for pointing that out. It was true that the size of our battery led to temperature constraints. But his concern was overblown. "You know," I pointed out, "ninety-nine percent of the time, the weather in North America is not tundra."

If Mr. Ice Age hadn't been so persistently negative, I learned later, Upfront would have increased their investment. Mark later told me that, as a senior partner, he thought it was important for him not to bang his fist on the table to get his way, even if he felt sure that a larger investment was wise. Return on investment is the name of the game, but camaraderie is also crucial for a successful operation. And if the senior partner simply overrules younger team members when he disagrees with them, even when their objections are dumb, then the juniors start wondering, *Why do I even work here? He's just the king.*

I could see why Mark had asked me the question about being a phenomenal CEO. He was particularly attuned to the importance of leadership.

After that meeting, Mark told me, he spoke to his associate. "Look, you grew up north and I grew up in California so you have a lot more insight about really cold weather. That's good. But entrepreneurs fix problems. They'll find a way around it."

Never mind that hiccup: There was a much more important benefit we got from Upfront than money. At the next gathering I gave them another progress report. I had hired a marketing/branding firm to do a survey, and maybe their most consistent finding was: "Spouses don't like the name DoorBot." I chose to call it "the spouse problem" and not put the onus on just one gender. We knew we had an issue with many couples where someone would excitedly buy a DoorBot and their spouse, before even seeing the product, blocked it just based on the name. I moved onto other subjects with Upfront: how we were thinking of changing our manufacturer in Asia; privacy; our forays into working with law enforcement. I discussed various metrics, and talked again about the ring of security that our product would provide, the ring of neighborhoods we were trying to create, the crime we were hoping to help prevent—

"Jamie, I have the name for your company," said Hamet Watt, one of the firm's partners, while tapping away on his phone. "I promise... hold on... I have it."

He turned his phone to me, hopeful. On his screen was the site Ring.com. The URL was listed for sale.

"Why don't you call it Ring?" said Hamet. "You keep saying 'ring'— ring of security, ring of safety, ring of neighborhoods... you said it like seven thousand times during the presentation. Just call it Ring. The domain's available."

Holy shit. I knew that Hamet used to be an entrepreneur himself, which always boosted a VC's standing in my eyes.[2] He was absolutely right. 10,000%.

I was stunned. I almost couldn't believe it. Ring. Ring.com. The sound of the doorbell mixed with our encircled concepts—security, safety, neighborhoods. Perfect. Musical. Timeless.

And so simple. One syllable. Four letters.

I could see from Hamet's search on his phone that Ring.com appeared to be owned by an individual, not a company. And the URL wasn't currently hosting business or any other content. I had to get in touch with the owner and negotiate.

Later on, Hamet texted me.

Got a guess on how much they want?

I texted back:

I would think it has to be $250k-$750k

Diego, who owned a bunch of domain names himself and had an unerring sense of pricing, thought it was worth $100,000, max.

I hoped he was right. Because whatever it cost, I had to have it.

O

I wanted the decisions I made about my company to be completely my own. That's harder than it sounds.

---

2   I make an obvious exception for the entrepreneur-turned-VC-turned-asshat who turned me down right before *Shark Tank* because I needed a "lead."

I know I'm going to fail. Like all entrepreneurs and inventors, I failed on a day-to-day basis, way more than I succeeded. That's in the contract. But I wanted to make sure that when I failed, it was *my* failure, no one else's. As Hamet would say later, "the single worst outcome for an entrepreneur is regret." (Honestly, that's probably true for everyone.) Failure sucks. But failure is better than regret, so long as it's an honest failure. If you have a point of view, a hypothesis, a dream, a mission, and you go hard for it in the way you believe you should, and you fail, it hurts. Bad. Maybe you go into a deep funk that takes a long time to escape, and you stop showering, and you watch bad TV and your loved ones start avoiding the very dark room you won't vacate. Eventually, you turn off the TV and clean up and go out into the sun and eat something healthy and put on your running shoes and give yourself the space to accept— forgive?—that you had the rare courage to follow your idea, hard. *Your* way. Yeah, it didn't work out the way you hoped. Sucks. That's life. But at least you left it all on the field.

On the other hand, if you have an idea, you pursue it, you believe you should turn left, but you let someone talk you into going right, and *then* you fail? That's *really* painful.

I would not let regret be the final takeaway. As long as I felt the correct path was to the left, I didn't care if every single person I knew was telling me to go right.

I got great advice and learned a ton about business from the venture capitalists who believed and invested in my company. I appreciated that, almost to a person, they wanted me to blow the roof off with my idea. They weren't in it for a 1.5x return. That's why I was so cost-conscious, even with smaller items: I told my team that every dollar of venture capital we spent was really $100, because we had taken that money to get them 100x what they'd given us.

I learned that hardware was even more brutal than I'd imagined. It takes a lot more capital because of up-front costs, available materials, reliance on the supply chain, something physical you have to pack and ship, visits to factories in rural Mexico or China, problems that cannot be fixed in a day the way software glitches often can, a one-and-done customer mentality, a harder-to-envision growth path, more and more rounds of money raising, more dilution, greater risk. I learned it's not smart to tell a VC who turned you down to go fuck himself, though sometimes it just feels good.

To be fair to VCs, I learned that it's hard for them not to get jaded, since they see hundreds of opportunities a year and fund just a few; spend so much of your time saying no and it's difficult not to look at things for what's wrong with them. (Correction: I learned that VCs never say no outright. There's zero value in it. Instead, they will "yes fuck you" (patent pending)—e.g., "It's very intriguing but not right for us now but please call us in six months and show us your progress"—even when there's no chance they'll write you a check in a thousand years.)

I learned so much more. There were all kinds of wisdom I could have absorbed and didn't. There were so many smart, good people who told me to zig, and zigging might have been the right move in the long run, but I was intent on zagging. Of all the things I would not bend on, this was #1 with a bullet:

Follow the mission. Follow the mission. Follow. The. Mission.

I was sure that that was our path, our only path, to success. I wanted to make our neighbors feel what Erin had felt when I first showed her my invention.

I hoped I was right. I'd been wrong many times before, and sometimes needed someone else to point it out. When I had started Unsubscribe, I needed funding and invited Sky Dayton, one of the most revered entrepreneurs and investors in Southern California (EarthLink, Boingo Wireless, eCompanies, LowerMyBills.com), to come to the Unsubscribe office to meet the team. He was underwhelmed. "A little hard to see how you monetize this," he said. I pitched him my long-term plans for the company, the valuable data we'd be gathering, and what we could do with it down the road.

Big whoop.

A few days later, Sky called. "Listen, I'm not sure about the idea," he said, "but I believe in you and I'm going to invest. Even if I don't think it's the right idea."

I thanked him, hung up, and stewed. I called him back a few days later. "About your comment the other day," I said. "I did a lot of soul searching. You're right. This *isn't* the right idea. I appreciate your faith in me, but I can't take your money." I told him I would find the best soft landing for the current business. "And *then* I'm going to find that idea."

"No one ever turns down money," said Sky, slightly shocked. "Every entrepreneur I know is maniacally focused on getting the next money. Go, go, go, go, almost robotically."

One true thing that binds venture capitalists and entrepreneurs: Our chosen fields are not for the faint of heart. At some point, you say yes or no, go right or left, zig or zag, then live with the consequences.

○

Fundraising is never-ending, as Sky had said. Could I keep raising money for DoorBot, which was soon to be Ring? Because if I couldn't, we were finished. I'm not one for celebrating a job being done, because it never is. With fundraising, it *really* never is. The instant the money shows up in your account, you'd better know the date and time of your next fundraising pitch. It's the money treadmill. It's a real boost when your friends believe in you enough to give you money or find you money. I was so thankful to Diego that he'd not only invested in Ring but persuaded Sky to do so, too.

Believe it or not, fundraising did not come naturally to me. I thought I was a pretty good storyteller, and with Ring I was lucky to have a good story to tell. But because my aspirations were not modest and I was shooting for the moon, a lot of my storytelling and picture-painting required my listeners to suspend disbelief. It was more "Imagine a world…" than "This is what we've found… " Given that I was pitching VCs, people wired to find what's wrong with something, it wasn't hard for them to reject funding me. They were simply rejecting a world that didn't yet exist, and might never. Hard to argue.

Sometimes I had to junk the "imagine this" approach and just go with bluntness. I was back east to pitch and sell in person, and took a detour to visit my friend and former professor Steve Spinelli, best teacher ever at Babson College, my alma mater in Massachusetts. Steve had made money as one of the founders of Jiffy Lube and been a savvy investor generally. He also could not help being my teacher and imparting guidance at any chance. I told Steve we were rebranding to Ring, building a much better product, planning to release it in the fall.

"Have you measured market demand?" he asked. "Do you have any proprietary technology? What's the channel of distribution?" Always the professor, he just kept digging. "Do you have manufacturing relationships?"

I smiled. "No offense, Steve, but I didn't come for advice. You want to invest, yes or no?"

CHAPTER 6

RING.COM

eBay.com. Half.com. Cars.com. Shop.com. Toys.com. And yes, Nest.com.

So many great four-letter domain names. And I wanted one: Ring.com.

The owner of the URL was willing to part with it... for two million bucks. That represented a massive chunk of the money my VCs were about to give me. Neither they, nor a couple of my seasoned tech friends who had experience with overpriced domain names,[3] thought it was a great use of my new capital. Nor did the fellow who ran the mezcal company on the other side of the wall of our Santa Monica office. "You're going out of business! Your doorbell doesn't work! It's just a name!" he yelled at me in the parking lot as I walked to my car one evening. On one hand, I wanted to yell back that he didn't know what he was talking about; on the other, I wondered if he was right and I was making a huge mistake. I also wondered where his anger at me was coming from, but realized he'd probably heard some of my own raging through the walls. "You're going to spend all that money on a stupid name?!" he yelled.

Another doubter wondered, "Jamie, does it really have to be four letters? What's so special about four letters?"

Yes, it had to be Ring. When I'd come up with my voice message-to-email transcription service, I first called it Simulscribe, and it stagnated. When I changed the name to PhoneTag.com, we got a burst of interest.

---

3 And sometimes incredibly cheap ones: Scott Marlette and his partners bought the URL GoodRx.com for $80.

Names matter. I had once thought they shouldn't—all that mattered was having a quality product with an easy-to-understand benefit, a great customer experience, and a fair price. Turns out, the name matters.

I would not make that mistake again with the doorbell. Soon enough, there would be lots of video-doorbell competitors whose products might be almost as good as ours when we launched F5. So the way to separate ourselves from the competition was brand.

A mission as big as reducing crime in neighborhoods deserved a brand. That brand deserved a great name.

For some totally unfathomable and fortunate reason, this URL owner showed zero curiosity about the individual or company that was trying to buy his name. In our exchanges, it seemed almost as if he was unfamiliar with the internet, which was particularly weird for someone who harvested domain names.

I got the sense that for some time he had overplayed his hand, consistently valuing the URL higher than the market did. Which happens. Maybe he had tried to sell it during the dot-com boom for $10 million and it was worth only five then. Or maybe *I* was the one being played, and he knew exactly how much a perfect four-letter domain name could fetch, certainly way more than I'd paid to own SlowDownAsshole.com ($15).

Diego urged me again, explicitly, to not pay a cent more than $100k for Ring.com. I explicitly did not tell him the owner's starting price.

First, I got the owner to knock the price down from $2 million to $1 million, but that was still an insane amount of up-front cash for a struggling startup to just light on fire, a full third of what I was getting from True. I had to figure a way to own the name without bankrupting our company... would the owner be interested in equity instead of cash?

No. Wow. Clearly he hadn't read about Google's recent multibillion-dollar acquisition of Nest. I made one last offer for slightly under $1 million.

Nope. One mill. We set a closing date.

I forgot one thing, though. I didn't have the money.

The morning of the closing, I called the owner. "Listen, I'm in the parking lot of my company and I'm so embarrassed. The bad news is my board won't let me buy the name, for the full price today, for what I previously offered you." It was not a lie. I had a board. The only detail I left out was that the board was just me.

"Wow," said the owner. "That's a dirty thing your board did."

"Tell me about it. Worse than dirty. Disgusting."

"I'm very upset."

"I hear you, brother. Me, too." I went on a bit. I doubled down about what a bunch of assholes my board were being. "But the good news is I'm authorized to deposit one hundred seventy-five thousand dollars in your account, *today*"—I had $187,000 in the bank; the VC investments had not yet closed—"and the additional eight hundred twenty-five thousand paid in installments over two years, for a total of one million dollars."

He lost his shit. He unleashed a string of four-letter words very different from Ring and eBay and Half. Effing this, mother-effing that. The connection dropped. He'd hung up.

Damn, I thought. Had I overplayed my hand?

Fifteen minutes later, I got an email from him.

Wire the money.

He included his bank information.

He never asked who was on the board. Never asked what we did. I hope I would have, in his shoes. Maybe when you're offered a million bucks overall, with $175k coming that day, you just want to get it over with as quickly as possible.

I called Adam at True to boast what a great deal I had cut, that I'd essentially just saved us so much money. He wasn't quite ready for high-fives; their investment was about to close, and already a significant chunk of it was gone because I had a jones for a great four-letter domain name. Adam was fully Team Siminoff but, as I'd done with others, I was not making it easy for him.

Ring.com. What a great sound. As sweet as the three-toned jingle the doorbell made.

○

Meanwhile, soon after that call, somewhere in a building just a block over from us in Santa Monica—the headquarters of Demand Media, a content company that operated online brands and owned a portfolio of domain names—someone was on a phone call with the company's recently departed chief financial officer, Mel Tang. They were just shooting the shit, and one of them said to the other, "Did you see that Ring.com just sold for a million bucks?"

"Holy shit. Who bought it?"

"I don't know. Some dude. Probably a jeweler."

"Bonkers."

"Yep. What an idiot."

○

We had to improve the design of the doorbell so that it looked more a part of the house. It needed to make a statement about what it was, yet not be obtrusive. It represented a safer home. But our first version had a bulbous camera lens that kept an eye on visitors and anyone outside the house, so we wanted to make the whole thing look less... sure, less creepy.

The more we could get the doorbell to look like house molding, the better. Our models were the planes and shapes found in a home—a doorknob, a banister. I needed more great design people. I already had

John Modestine, an amazing product designer who could seamlessly integrate the work of electrical engineers and mechanical engineers. I needed another talented mechanical engineer, so I called my cousin Mark Siminoff, who had valuable experience in manufacturing and plastic injection molding, to ask him to come work for us. He wasn't biting, not at first: He had a full-time job running EarthBaby, a company he'd cofounded in the Bay Area that did good for the world by collecting single-use dirty diapers and turning them into commercial compost. Since I never give up, I did not accept Mark's no. When I flew up north to plead with him to give me at least *some* time to assist with the new design, I often had to visit him at his diaper company, or "the warehouse of shit," as I lovingly called it. (Maybe 1523 wasn't so bad after all.)

Eventually I intrigued, or exhausted, Mark enough that he agreed to help. I didn't have the money to pay him anywhere near his value. But he was my cousin, so I got a family discount.

Mark recruited Chris Loew, a top industrial designer with whom he'd worked at IDEO in Silicon Valley, to address our design problems. Chris called the previous design "gadgety" and sketched out a thinner, sleeker version that would sit as close to flush as possible against most housefronts.

Sleeker as version F5 looked compared to bulky DoorBot, I pushed the design team to make it even thinner. As Chris said, "Whoever gets the device closest to paper-thin, wins." I wanted the doorbell to measure 21 millimeters from the wall to the face of the device. The biggest task would fall to our engineers to make the 30-odd components plus battery fit inside the tighter space. Chris came up with design concepts, Tim Simons worked on the PCB (printed circuit board) layout, and they would hand their drawings to Mark to implement them in SolidWorks (computer-aided design (CAD) software). It reminded us of the challenge of the Apollo 13 rescue, where Mission Control had to figure out everything that wasn't essential and could be eliminated, and keep only what was absolutely needed for a successful re-entry, all inside a much more compressed space. We were doing that while also trying to make our next doorbell look Apple-Dyson-Samsung-level handsome. Failure was not an option.

One thing I wanted was perfectly even illumination of the lighted ring around the button. We used 16 LEDs to encircle a "light pipe," which acted like a prism underneath to create an even 360-degree lighting pattern. And we had to do that while being able to draw enough power for the LEDs to light the ring brightly enough to be seen from a distance.

Then there was a challenge so advanced, no one had yet conquered it (or maybe even attempted it). Our doorbell needed to do F5-worthy motion detection, and this would be the first battery-based product capable of ultra-low-power motion detection. How did we intend to pull it off? By using heat zones. We mounted passive infrared (PIR) sensors inside the doorbell. In front of them we placed a Fresnel lens that matched with the sensors, allowing it to detect movement through heat energy: The heat of a person near the doorbell would be detected on the surface of the PIR sensor, each of which could be only on or off. If you had only one sensor, it might get triggered by motion you didn't want to trigger it—if, say, there was a busy sidewalk to the right of your home. By placing sensors at angles to one another and overlaying them, we created multiple zones; you could choose which zones to be sensitive to changes and which not.

To make this work correctly was difficult for a bunch of reasons. For one, the sensors needed to move, so they required extra space inside the housing (which meant eliminating space elsewhere). And the Fresnel lens had to be a precise thickness because it would be injection-molded in two pieces to make the lensing effect work. And the lens had to be made of different plastic from the rest of the doorbell housing because it required specific light-transfer properties. And we couldn't just glue the lens and doorbell housing to each other, because they were different types of plastic. And we needed a waterproof seal between the Fresnel lens and the housing.

Miraculously, we got the doorbell, with all that going inside, down to 21 incredible millimeters thin. We thought we had accomplished something as monumental as what NASA Mission Control had done with Apollo 13. Tom Hanks is playing the Fresnel lens in the movie.

The whole team was heroic. I had pressed them hard, because we were running out of cash and I had promised Jon Callaghan at True, and our other investors, that we would deliver by October 1. One of our engineers, Andy, worked truly ridiculous hours to get the doorbell done and worthy of F5 reviews. I thought about my promise to Jon and how I knew that the only way we could pull off a whole new, great product, pretty much from scratch, in the space of nine months, was to "break" some of my team members. Andy seemed to be at the office 24 hours a day, 7 days a week, for 6 months, pushing himself to the breaking point.

With the design all done on our end, I needed to hand the CAD model and drawings to the manufacturer but first—just to make things even *more* complicated—we needed to find a new factory. It wasn't because Tatung had created that idiotic antenna-shaped-like-their-logo problem; after all, my deciding to sheath the doorbell in gorgeous metal had been the real issue there. But every time I contacted Tatung about moving forward on F5, this amazing, superior video doorbell for them to build, they were noncommittal. Things dragged on, without clarity. It turns out that factory owners, like venture capitalists, also see zero value in saying no outright. I finally realized that, for whatever reason, they did not want to be in the doorbell business with us. Maybe they thought that after something as faulty as DoorBot, we were not capable of a truly great product. Maybe my blowing up at them for the logo-antenna fiasco pissed them off. Maybe they felt bad that rejecting us outright would kill us off. I couldn't wait any longer to get the answer. I needed a new factory.

I contacted everyone I knew with manufacturing experience. An acquaintance I'd once made who worked at Foxconn, the massive, Taiwan-based electronics manufacturer with a huge factory in Shenzhen, China, had once told me that if I was ever in his part of the country, we should meet for coffee. Now I called and told him I was going to be there in a few days on other business (totally not true) and was he around? *Yes? Great! See you in a few!* I booked a way-too-expensive flight to Hong Kong, flew the 16 hours, then took the train to the border of Shenzhen, a city that links Hong Kong to mainland China and has twice the population of New York City. I checked into a little hotel, showered, dressed, and waited for my acquaintance to pick me up for coffee.

He called, deeply apologetic: Coffee was off. Terry Gou, head of Foxconn, needed him and his team to fly right away to Korea to visit a big customer. "Can we do it another time?" my contact wanted to know, under the totally reasonable, mistaken assumption that I was there "all the time" and currently on "other business."

I acted brave, tough, indifferent—"No problem! Yeah, we'll pick it up next time! All good!"—but inside I was dying. Fortunately, my acquaintance was sufficiently contrite that he wondered if I might want a tour of the Foxconn factory, with one of his people?

"You know what? I've already set aside the time for you... sure!"

Soon after, a car arrived in front of my hotel, a Mercedes stretch limo with the license plate FOXCONN. I convinced myself it was Terry Gou's special car. At the factory, I met Chien Lin, a young Taiwanese American who'd spent his middle adolescence through his college years in Texas, so his English was impeccable. Chien and I hit it off immediately and he loved what we were trying to do at Ring. *How could Foxconn help?* he wanted to know. It felt as if we were at similar points in our career, though Chien was younger: I was a nobody trying to make a disruptive product; he was a nobody trying to do something disruptive at Foxconn. We were both go-getters, and he had an idea: Foxconn was talking about creating a division, to be called Foxconn Labs, dedicated to startups. It was a great way for them to meet high-quality entrepreneurs with great product ideas and build early relationships. Why not make Ring the trial balloon? To help, they would give us 120 days from the time our product got to the US before we had to pay for our order. In return, they would get 5% of the company.

It was almost too good to be true. Except for Arrow Electronics, which was giving us similarly great terms on chips, all of our other vendors required that we pay them right away, totally standard for a company of our size. Or we'd have to pay a percentage of the total cost when we ordered the components, the rest when the parts left their facility. That's how it was with Tatung. If we were very lucky, a vendor might give us 30 days, once in a while 60. But a whole 120 days? Given how fast we burned through money, those four months were a lifetime.

To celebrate our impending partnership, Chien and I went to lunch, where I learned it was hairy-crab season. Not a typo. Google it. One of my least favorite three-word phrases. A delicacy often flown in from Yangcheng Lake in the Jiangsu province, they can cost up to hundreds of dollars per crab, and I guess if you like eating shark fin, it's not a big leap to hairy crab. Or maybe it is. My server opened up the center of the crab for me, exposing organs and goo. *Holy fuck*. This was an incredible journey that Chien and I were about to embark on, so I did not want to insult him or the others around the table but... *holy fuck*. I searched for anything in the crab's body that looked even remotely edible—and there it was, finally: a glistening chunk of white meat. I dug in with my chopsticks, pulled out the nugget, and just as I was directing it toward my mouth, Chien slapped it out of my grasp.

"Don't! The lungs. The stuff inside it is what they pick up from the sea floor. It has bones in it." Apparently, there were contaminated parasites and potential toxins in there, too.

Chien did not make me eat the rest of it. Or any of it. Foxconn loved the Foxconn Labs idea and approved the terms with little hassle.

I got the deal because Chien was a maverick. He wanted to see us thrive almost as much as we did—hence the favorable terms that gave us the chance to breathe.

One entrepreneur understanding and helping another. I had my factory.

O

To build the brand, we also needed to make ourselves, the team behind Ring, appear as responsive as possible. I included my email address on every box. To do social media and general branding, I hired Yassi Yarger, another one who'd answered a Craigslist ad. She was 21, had studied English lit and communications in college, and had been working at a PR agency for a few months. I liked her hunger. In the interview I asked her to name her favorite tech product and she said her iPhone, and I told her that was a great answer because Apple products were the kind we aspired to make. From the moment she started with us, she was doing all kinds of things besides PR, everything from helping with fulfillment, to keeping track of which competent team member I'd fired in a moment of intensity/insanity and needed to rehire. And everything in between. But she, like August, also had to deal with pissed-off customers. When she logged onto the old DoorBot Facebook page, she was bombarded by roughly 5,000 unread messages, dating from months earlier. All of us, including me, had to pitch in and respond to them, one by one, by email or phone. A couple of weeks in, I wondered if Yassi was having second thoughts about joining the company, but I also knew that this was the best crash course in business she could ever get. Because to respond well to angry customers, you had to understand the mechanics of the product, the firmware issues, when to tell a customer to reflash their device, why, and hundreds of other pieces of know-how. To advise customers who had dropped $200 on your product only to see it act up within hours of installation, you needed to know *everything*.

The broad response I had the team give to neighbors—"Yes, DoorBot has some issues and we're working through them as quickly as possible"—was tough to sell. Correspondence with customers was frequently filled with four-letter words, not the fun kind. In ALL CAPS. Yassi looked frustrated.

"Just tell them it's coming," I told her. "We're one firmware update away." I kept moving the goalposts.

Yassi wanted to send DoorBots to the press for reviews.

"Nope," I told her. "The product does not work well enough." Even with the miracle on Christmas Eve-Eve, we still had lots of issues.

"Okay... then what do you want me to do?"

I had no good answer. "We're all just trying to keep the boat afloat, however we can." Working at Ring had become boot camp for scrappers.

O

By mid-summer 2014 we'd sold roughly $4 million of the first-generation DoorBot product, which we would stop selling at the end of August. We weren't making any more of them. We were spending almost all of our resources and time on the second-generation doorbell product, F5 to us, the Ring Video Doorbell to the world when we eventually launched. We had final electrical and mechanical designs locked down. Through True Ventures, I'd met the Fitbit team, which helped us optimize the wifi chip and the Ring prototype. Our design was so much more sophisticated— just way better—than DoorBot.

I remained quietly, pleasantly surprised that people were still buying DoorBots. It was a case of our filling a need (turns out people love the idea of a wifi video doorbell!), and not being very good at it (yet), but no one filling it better. We sold most of our existing product at GetDoorBot.com and Amazon, and I wanted to drive initial sales for our next product mostly online, too, this time with an all-time great domain name, Ring.com.

We were in conversation with Home Shopping Network (HSN) for me to appear when the new doorbell launched. The margins selling through HSN or QVC were famously low, but the volume they moved was famously high. Plus, I could test different versions of the pitch to see which ones hit best, since HSN (and QVC) tracked calls and purchase volume *as* you presented, meaning you got a real-time thumbs-up or thumbs-down on your message.

Yassi was writing back to all of our neighbors who were leaving 1-star reviews on Facebook and Amazon, responding to them publicly, since that was our billboard. People needed to know that we took their complaints seriously, didn't ignore them, and most importantly were doing something to correct the problems. I got on planes to fly to the cities of commenters who'd given us detailed one-star reviews to reinstall their DoorBots myself; I was able to make most of them work and turn irritated customers into 5-star neighbors. It was not a scalable exercise, but consumers needed to know we meant what we said about caring for them.

I was in talks with retailers—Best Buy, Home Depot, Brookstone—that had expressed interest in carrying the new product for the holidays but got no real traction. Maybe the stench of DoorBot had them nervous.

We hired two designers in their twenties from Pasadena to help with our color scheme and logo, but I found it a waste of money. They didn't tell us anything we didn't already know. August and John worked on the website and used burnt orange as their main color. I didn't like it; I didn't think it invited engagement. "Facebook blue" seemed perfect. I told the Siminoff Brothers that Scott Marlette was coming by the office to look things over and give his opinion.

"Yeah, I wonder what Scott, employee number twenty at Facebook, is going to say," said August, achieving a whole new level of eye-rolling.

Scott looked at it. "Blue," he said.

I enlisted my friend Karni Baghdikian, a college buddy and later my roommate in a shoebox studio apartment on the Upper West Side of Manhattan, to direct and produce several launch videos for us. Karni and I had been in New York City on 9/11 together, a bond that never left us, and he had since grown into a brilliant designer of websites, videos, and short films. He'd done a viral video campaign for Carl's Jr., and digital campaigns for Kia and Bacardi. As Jersey boys, we saw the world similarly, as something you had to go out and take; not surprising for him, since he'd made collegiate rugby All-American at 5'6".

We used my front door for the video. Karni's friend and his wife's cousin played a couple in bed when the doorbell rings. The couple sees on their smartphone that it's the nosy parents (played by Karni's mom and dad) of the woman in bed. Another video showed a woman (actor) playing with her child (one of our neighbors) in the park, and "answering" her door from there. Karni attacked our videos like they were Oscar contenders. I started calling him "Karni Scorsese."

We were busting at the seams. We used racks as desks. Four out of five new people werecustomer service, and they'd basically taken over the warehouse. Every morning someone from shipping used the forklift to move all the boxes of product outside. Whatever wasn't sent out that day got moved back inside at night. Had I been an entrepreneur in Chicago, with its more challenging weather, rather than LA, I would have had to devise a different plan.

I flew to Seattle, where I was introduced to an executive at Amazon, Nick Komorous. I brought a prototype of the F5 to show him, a sneak preview. Few people had seen it outside of the team building it. It couldn't hurt to have someone at Amazon excited for our direction.

Chaos, action, energy, attention, progress. All good, right?

O

With so many ups and downs on a daily and even hourly basis, I had to guard against "the trough of sorrow." That's what Paul Graham, co-founder of the startup accelerator Y Combinator, calls it.

The trough of sorrow usually comes in the middle of the startup journey. If you graphed the whole thing, the journey would start at the top: *Great idea! Look at all those users! Good buzz! Real money coming in!*

After a while, though, you're hit with reality. Returns. Customer-service calls. Zero-star reviews. Some bad apples among your team. Morale slump and burnout. Financing issues. All of it can happen to anyone. It *does* happen to everyone. After all the work, all the hours, all the intention to build something real, something that gets bigger and better each day, you find yourself in the trough of sorrow.

Oh, fuck.

On the graph it looks like squiggles of disillusionment. It's not quite down to zero— you've built something good, after all, and it's got some intrinsic value—but these are the dark clouds you've been warned about.

After a while—if you keep your head down, stick with it, keep believing, and get your team to join you—you start to look up at the sky. You poke your head out of the trough, see new sparks, acquire more users, and achieve that holiest of holy grails: a lower cost of capital. You emerge, if cautiously, from the trough of sorrow. You hope never to inhabit it again.

You never know quite when it's going to get you. But I had entered the trough of sorrow.

I was ready to sell the whole damn thing, but no one would buy this liability.

It was one thing piling on another. I couldn't let the worst of the bad DoorBot reviews go without responding like a madman. ("Get better wifi!") I was terribly underslept. I had already made a bunch of costly decisions, not to mention pissed off a number of people I cared about. Was it worth it? The whole thing was taxing emotionally—and we were just two years into the journey, one that, if you're lucky, usually takes 7 to 10 years. I still believed in the great mission of ours, that we were there to help reduce crime in neighborhoods. But I was fried.

Fuck it. Enough. I just needed to make back enough money for my investors, and, hopefully, a little more for myself and the rest of us.

Through a friend, I scheduled a phone call with Naren Gursahaney, the CEO of ADT, the biggest player in the industry, by far. But wait—what good was a phone call? What about my rule always to be face-to-face, if at all possible?

So I just went. ADT's headquarters were in Boca Raton, Florida. I bought a plane ticket to Fort Lauderdale, flew cross-country, rented a car, and arrived at their office 15 minutes before the scheduled phone call. I told reception I was there for my meeting with Mr. Gursahaney.

A few minutes later, Naren's assistant popped out, eyes wide with panic. After introducing herself, she asked, "Oh my God, did you come in for this?"

"Yep, just flew in!" I said cheerily.

"Oh, no. This was scheduled as a *phone call*, not in person—"

"Oh my God," I echoed. "No, no, *I'm* the idiot. Listen, I'm so sorry, let me go to the parking lot and get in my rental car and do the call from there..."

She said I should come up to his office, that the CEO was in. I'd stormed the ADT castle, but it was perfectly possible that Naren might have planned to take our call from elsewhere. Lucky for me. You miss 100% of the balls you don't swing at, and I was glad I'd connected on this one.

In the world of home security, ADT was Coke and Pepsi, McDonald's and Burger King and Wendy's all in one. Rather than play it coy, I shared with Naren our whole playbook. Our mission. I even shared the particulars of the inner workings of the new doorbell. I showed him the most recent version of the Ring, which so far had been seen only by me, the team working on it, and Nick at Amazon. I vomited out everything and anything I thought he might find relevant or impressive. Why? Because I wanted him to buy us! *Just make me an offer, right here, **please?*** I imagined my mind making his lips move to form the words, "We'd like to buy you for ten million, Jamie. How's that sound?" With that amount, I could just about pay back my investors and have a little money left over for me, and we'd all go work for ADT.

Naren did not make an offer. It was *Shark Tank* minus Mr. Wonderful.

I was so disappointed. I'd gone on *Shark Tank* to get a real offer and walked away disappointed. Now I'd flown to Florida, hoping my biggest future competitor would buy us and maybe put me out of my misery, and I was disappointed once again.

On the drive back to Fort Lauderdale Airport, part of me was happy that I had failed at my goal. I loved our independence. But I think I did what I did because what I wanted most—to accomplish our mission of reducing crime in neighborhoods—was seeming so difficult with our budget and timetable. Why not find someone who shared our commitment to home security and had a way bigger war chest?

But something else was at work, as I wallowed in the trough of entrepreneurial sorrow. When I look back on what was going through my head at the time, I think I was sick of the truth-stretching. I don't mean the storytelling; I was good at that. We had a great story to tell, a story about how an innovative doorbell, born of the simple need of an inventor in his garage needing to hear visitors and package deliveries at the front door of his house, could make neighborhoods safer. That's an incredible story. Yeah, that's not what I mean.

No, I mean the truth-stretching. The reality-distorting. I do *not* mean lying. I mean the posture you assume to achieve goals that others might not consider achievable. I mean those things you say to close the deal, raise the funds, recruit the partner, get a good spot on the retail shelf. I mean the promises you make to those who work for you and those whose money you rely on.

Stretching the truth is what entrepreneurs do. Maybe it's what anyone does who wants to achieve something great. We push and we push. We distort reality. When I think back on my mindset then, I can't help but think of Elizabeth Holmes and Theranos, the health-tech unicorn that wasn't. She was the company's founder and leader, and she really, truly looked as if she might disrupt an entire industry, one of the biggest in the world. Instead, she went down in flames ignominiously. She was indicted in 2018, convicted in 2022, and went to prison in 2023 for overstating her company's capabilities to investors.

I'm not defending what she did, especially given the stakes: administering diagnostic tests that supposedly told you almost instantly if you did or did not have certain disease markers or the disease itself. But I'm also not passing judgment on Holmes. I don't know all the facts.

No, her problem might have been picking the wrong industry.

Almost every entrepreneur has to stretch the truth, to themselves and to everyone around them who they also need to buy in if they're to reach the promised land. Every entrepreneur is constantly hallucinating—though actually they're not. Is it misleading, deceiving, even lying if I truly believe I can make something happen, even if no one around me believes it?

If you lie about blood tests, then yeah, that's a real problem. But what about when the doorbell you ship doesn't work? I mean, it worked... but not always. And when it did, not always well. In fact, often not well. Give me a pristine wifi signal and a house made of balsa wood and yeah, sure, my doorbell worked perfectly. Or at least the second batch did, usually. But what about the real world? Does that make me and every other entrepreneur a fraud?

The product and the promise matter. A doorbell is a doorbell. If it fails and does not live up to the promise, you can get your money back. While we were getting yelled at constantly and handling a flood of 1-star reviews, the customer could always return it, something people took us up on less than you'd think for such a challenged product. A lot of great innovations—almost all of them, in fact—launch with a serious problem of one kind or another. It's okay to launch with problems. That's how it works with technology. That's how you learn. That's how you get better products.

The companies out there that make things fine the first time are usually the ones that fail to build great long-term businesses, because they picked a problem that was too easy to solve. Those companies usually die quickly, since everyone else has access to the same technology they used to make that small, perfect, easy thing. They broke no glass. Probably they punched no holes in walls. But companies that have problems with their first products—real, substantial, existential problems, like the ones we went through again and again (though ours never had anything to do with customer safety)—*they're* the ones that tend to become big because they get used to solving hard problems. Reid Hoffman's great line is worth repeating: "If you aren't embarrassed by the first version of your product, you shipped too late." Those are the companies breaking glass. Maybe they—we—stretched the truth because that was the only way to achieve something no one thought possible.

I wish for all of humanity that Elizabeth Holmes had decided to start in something less critical; my guess is she would have had a long-term positive impact on the world if she had.

Am I saying that we entrepreneurs who swing for the fences are no different from Bernie Madoff and everyone who ever ran a Ponzi or pyramid scheme? No, we're different. Because those people are con artists who go *in* not believing things themselves and just hoping everyone who hears them is a sucker who believes what they don't and never would. Their reality distortion never leads to a better world.

With me, with most entrepreneurs: We *believe!* We want you to believe also—we *need* you to—but the key is, we're right there with you. We're not going away. Belief is a powerful drug. We'll go down with the ship if necessary.

Every entrepreneur does this reality-bending, often, because we have to. To VCs, customers, media, industry analysts, our team, ourselves.

And one more thing: Is it lying, bending reality, hallucinating, truth-stretching... if the crazy, improbable thing you promise eventually comes true?

O

As we scrambled to make sure that we could announce our new and improved Ring doorbell by October 1, I got a call from our *Shark Tank* producer. The show had heard about all the orders we'd gotten, the money we'd raised, the buzz, the rebrand, and wanted to know if I would come on for a follow-up appearance.

Are you kidding? Of course!

The show had done this only once before. But given the attention my rejection of Mr. Wonderful's offer had gotten, and what a nerve the product had struck with the public after we'd aired, the *Shark Tank* producers thought it would make for great TV. But we had only two weeks to prep for the taping.

We would make it work.

I couldn't believe our luck. The first appearance had just about saved us. We still needed money; hopefully this time we would close a deal. Given our upcoming relaunch with a way better product, this follow-up appearance could skyrocket us. I was relieved that ADT hadn't made an offer to buy us, and that my mind did not have the power to move their CEO's lips.

I told Dave to unbury the home façade from my first appearance on *Shark Tank*. We would repaint it so it sparkled for the next appearance, blue with orange trim, Ring's new colors. In the meantime, I rehearsed my update, which would be a lot easier since I'd given so many pitches in the last year and because there was now such a strong narrative around Ring. I wanted to tell the Sharks that their reaction a year earlier was the reason we had reinvented the product. I wanted to thank them for that and give them another bite at the apple. I wondered what Mr. Wonderful would say to that.

What a reversal of fortune since that horrible Christmas Eve-Eve. *This is why you hustle*, I told myself, *even when you don't want to*. Because the more you do, the more lottery tickets you accumulate, the likelier you are to get what you want.

The night before I was to appear on the show again, Dave and I were in the back of 1523, among thousands of boxes of doorbells, putting a fresh coat of Ring blue on the housefront. There was a phone call. A number I didn't recognize.

Someone from *Shark Tank*. The show simply didn't have enough time to make it work. *Really, really sorry not to have you on again, Jamie*, he said. *But good luck!*

Weeks before our relaunch, the greatest free commercial for Ring was not going to happen. The Ring bank account would be back to its most natural state, empty.

I stood there. Dave looked at me. Once again, I felt as if I might cry. This time, I did.

O

Ollie started at a new school, and Erin and I were invited to a welcome party by parents of one of his classmates. As we approached the house, I saw a DoorBot installed next to their front door.

"Shit," I said under my breath. "There's no way that thing works."

"Should we try it?" said Erin.

Maybe it was a brick. Maybe one of the later ones we fixed. Then again, even if the camera worked great, it might have wifi problems. Or audio problems.

I knocked.

Inside, we were greeted warmly by other parents of Ollie's classmates. We knew no one. I got myself a drink. The owner of the house came up to introduce himself. He said he knew who I was, what I did, what I was responsible for. "Listen," I said, "I'm sorry—"

"Sorry? What for?! Are you kidding? I love the idea! Yeah, I've had some problems with this model. But the better one's coming out soon, right?"

It struck me then how lucky I was. Not just because there were people out there with forgiving natures. But because ours was a product that people *wanted* to work. *Needed* to work. Everyone who'd seen my appearance on *Shark Tank* and went ahead and ordered a DoorBot was *rooting* for us. Why not? We were the underdog. However small our product, our mission wasn't. So many of the people who bought our video doorbell saw it the way Erin had the first time I showed it to her: *This makes me feel safer.* We would never have been granted such charity had we just been working on a more convenient or more beautiful version of something that already existed. Or something that was really cool yet inessential.

But safety? Security? Your family? Your home?

In the weeks leading up to the announcement of the Ring Video Doorbell, I went to bed terrified every night that I would wake to a Twitter frenzy that Nest or Dropcam or some even scrappier startup than us had launched a video doorbell. And every day that passed when that didn't happen, I would think, *We're so close.* I was—as I so often felt—a barbell of confidence and fear.

Diego, ever bullish on Ring's prospects, bet me my beloved Land Rover Defender, which he also loved, that we would do $30 million in sales within a year. I thought that was far too optimistic. I didn't think I was in jeopardy of losing any vehicles. I wouldn't have minded if I did, of course. (Also, just for the record, Diego, we never established what *I* would win if we *didn't* hit $30 million.)

Then again, a real entrepreneur never, ever bets against himself.

O

We broke glass. We broke gravity. In just nine months, we had created a revamped video doorbell with proprietary motion alerting, announcing it on September 29, two days before the date I had promised Jon Callaghan at True Ventures.

Unfortunately, we also broke some people, casualties of our war, including a few brilliant engineers who never seemed to leave their desks and who solved so many of the most difficult technical issues. If I ever forgot how hard I had pushed some people, the team didn't let me. For my birthday, they presented me with a punching bag, with photo printouts of many of their faces taped to it, weighted by who had incurred the most wrath. Mark Dillon's face took up half the bag. August's and John's faces were prominent, too.

The day we announced the Ring Video Doorbell and the start of pre-orders was a pivotal day in our company's history—and I wasn't there.

I was too nervous to be in the middle of it. I booked a trip to be out of the country.

I flew to Montreal with some local businesspeople from LA. It didn't even matter why they were going or where; I just needed to be away. The people I met in Canada learned that my company was having a moment, a rebrand, and for the whole day they dubbed me "Ding Dong." They outdid one another with doorbell jokes. I'm surprised they didn't serenade me with "Frère Jacques" and its famous bell-ringing. They were nice enough, but clearly thought my business was trivial, halfway to a punchline. Then, when they saw me checking my phone repeatedly, they started to feel bad about all the ribbing.

"Do you have any numbers yet?"

I did. By evening, we were above even our most optimistic pre-sales projections. *Way* above.

Within weeks, we were shipping. Within days after that, the effing 5-star reviews started rolling in.

We had broken glass, gravity, a few engineers, precedent. Maybe now, for once, we wouldn't be broke, period.

We had launched Ring.

CHAPTER 7

I BROKE IT, I FIXED IT

Growing up in New Jersey I loved radio-controlled cars (the Clod Buster and Frog were favorites) and BMX bikes, but even more I loved how often they broke because that meant they could be fixed. I tinkered in the basement with my cars or built things from parts bought at Radio Shack or Hobby Shop in downtown Chester. I'm not saying that means there was something broken in me that needed fixing; I just loved MacGyvering my way through life. If something didn't work, I taught myself how to fix it. I taught myself to fabricate tools. It's how I learned engineering and electronics. "When something works, it's less challenging, it's less interesting," the legendary inventor James Dyson once said. "If something's gone wrong, you want to know why it's gone wrong, and it's a learning process."

I was always moving, going, heading somewhere—riding my bike through the woods, or the miles and miles to friends' homes, or blowing things up outdoors, or crashing skateboards or motorcycles, sometimes breaking bones. I never stopped. If it sounds unhinged: I didn't do drugs. I didn't cut school. I was convicted of no felonies. Today, they'd call that "hyperactive."

My parents gave me what I needed—my father warm and understanding, my mother tough and efficient—but I was not a happy kid. I had friends but I was not part of any group. Maybe you figured that out from the "I tinkered in the basement" part. I was not a good athlete. I tried lacrosse and baseball but sucked at both, so I ran cross-country. Distance running didn't demand skill so much as suffering, and I could take pain. It was one place I felt I had an edge.

Schoolwork, same story. I drifted through classes, distracted and restless, wrestling with undiagnosed ADHD and the kind of dyslexia that turned windows into escape routes. Every now and then a talent broke through, like the time I designed a dream estate in an architecture class, a fantasy made physical in miniature. It was a glimpse of what I wanted but couldn't quite reach.

I wanted something I didn't have. Zach Vella, my best friend since we were 4 years old in pre-K, was better at getting things he wanted. Somehow, when we were kids and then young teens, he managed to live like Gatsby, though we didn't have that kind of money. More than once I heard Zach say, "Can you break a hundred?" and the store owner shook his head, and I was left shelling out whatever the thing cost. This was Chester, New Jersey. *No* one had a hundred. That takes talent.

When I was a junior in high school, my dad and I went to a local car show that featured the Land Rover Defender 90, a British-made off-road car and the single coolest thing I had ever laid eyes on. It wasn't just a car; it was freedom, rebellion, arrival. It was the vehicle of my dreams, the thing that said I belonged, I measured up.

"You want one?" my father asked.

I stared at him. Had he really just said what I thought? Maybe my dad was cooler than I knew. I nodded.

"Get all A's and it's yours."

I worked harder than I ever had, grinding through every class. When the report card came, I had straight A's... except for one stubborn B in algebra.

The sting of *almost*. The ache of falling short after giving everything. I could have pushed harder, run longer, found other ways—

Because my dad was so fair and decent, he looked up after examining my report card and said, "Okay, one more shot. Do it this time and the deal still stands."

I aced everything, including algebra, and he kept his part of the deal.[4]

I knew that earning my own money equaled greater independence. I took odd jobs, summer jobs, weekend jobs. I did landscape maintenance (weed-picking in the summer, shoveling in the winter). I painted houses. I was a bellboy at the local hotel. I worked with roughnecks (loved it) who ate their lunch with Popeye arms and greasy hands, and sometimes after a 12-hour shift I had to go to school. I shoveled horseshit from stalls of neighbors' barns on the way to school. I was always hustling.

When the richest man in my hometown, whose son was a friend, asked me to list the colleges and universities I was considering, I named Babson College, which I knew nothing about except that the high school guidance counselor had mentioned it (among many others). My friend's dad said, "So many of the most interesting and successful people I know went there." Done. Thanks to my father's creative way of motivating me in school, I was able to earn a spot.

Babson, a business college in Wellesley, Massachusetts, was (and is) a great, scrappy school, always punching above its weight. It's the size of most people's high schools. It's got a pretty campus. Everyone there thought they were an entrepreneur, and many of us actually were. But, as I wrote earlier, I wanted to be *the guy*. I made money doing homework for other students, especially the foreign ones who needed help expressing themselves in English; my compensation was Estefano letting me drive his Porsche 911 Carrera 4S convertible, stick. My buddy Matt Seney and I put up posters advertising our services: For $10 an hour we would do anything, Taskrabbit before Taskrabbit. When the job was particularly unpleasant, we'd find local high schoolers to take on the job for $8 an hour and pocket the difference. My friend Ronit Levy and I started a company called Gadgetronics, which sold (this is not a quiz) gadgets and

---

4   My poor brother, John: He was way better in school, smarter, better at testing, so he never got such an offer.

electronics. For me, this wasn't yet the start of entrepreneurialism but straight-up hustling.

In Babson's renowned Foundations of Management and Entrepreneurship (FME) class, where everyone starts their own legit business, Ronit and I planned a coffee shop, but so did another pair of students. (I know, sounds like an episode of *Curb Your Enthusiasm*.) We opened before they did, but their compelling ad campaign was the talk of the campus. "IS HE COMING?" asked one, with its religious implication hardly veiled. "HE IS COMING," shouted another. All over campus. Ronit and I wondered: *Do we really need to come up with our own marketing campaign?* We printed posters that simply read, "To our café...," with directions to our coffee shop, and taped them underneath each of their posters. Everyone got mad at us—the professor, our rivals, fellow students—and, in the spirit of their campaign, we offered contrition. But wasn't that a classic guerrilla-marketing stunt? Shouldn't they have *applauded* our cleverness? We outsold our competitor, maybe because of the attention. It taught me that all press is good press. It taught me (I hope it taught our competitors, too) to never leave your flank open.

My favorite professor was Steve Spinelli, who taught franchising and entrepreneurship. He was something of a legend at the school, not just because he had co-founded Jiffy Lube but because his classroom discussions were always more than just business nuts and bolts. He was interested in the humanity of business, finding the balance of the quantitative and the qualitative, the sweet spot between EQ and IQ. He loved to shoot the shit in office hours as much as his students did. He was generous with his time and insights. Once, I gave him a lift down to New York in a snowstorm, and since he was a captive audience for roughly eight hours, I must have asked him 500 questions. When I dropped Steve off in midtown, he said, with a smile, "God, you're exhausting."

Senior year, my friend Dan Drabinsky and I won the Muller Prize, the Super Bowl for Babson students. It was a competition for the best business plan. Dan and I conceived BSF: Big Storage Facility, a modular self-storage business where a truck brings a shipping container to you, you fill it with possessions you won't be touching for a long time, the truck returns to the facility. What made it better than places like U-Haul? One, you could stack the containers (or pods), so you saved on facility square footage; two, people didn't access the facility as frequently as with other storage places, so you cut down on the need for hallways and non-freight elevators, which saved more space and cost.[5] Dan and I split the $10,000 first prize and spent almost all of it on beer for the next few campus-wide parties. No one got mad at us.

At Babson I learned the nuances of market demand, how to manage a business dynamically (in theory, anyway), the pros and cons of business models, so much more. Still, for me, it was all about hustling. Work harder, accumulate more lottery tickets, earn more chances to win.

Dad gave me a watch for my college graduation, but not just any watch. He'd asked what I wanted and I said a stainless-steel Rolex Submariner. It was the one truly distinctive thing that lots of seemingly successful people wore, the one thing I wanted to show that I was the guy. Or at least on my way to being the guy.

After graduation, a friend of a friend called to say he was raising money for a hot dog chain in Hoboken, and could I write a business plan? I was busy looking for a job so I told him I couldn't. Which made him more eager to hire me.

"Okay, but if you *were* available, how much would you charge?"

Joking, I said, "Ten thousand dollars."

When he said, with no hesitation, "Yeah, that's fine. So just let me know when you free up."

It took a few seconds for me to realize someone had just agreed to pay me $10,000 for a job I had no real interest to do.[6] I started Your First Step,

---

5   A friend in New Jersey with a warehouse tried it.
6   If you're thinking, "Today, with ChatGPT...": You're right.

a business-plan company. After the hot dog chain plan, I put together a proposal for a robotic car wash, a capital-intensive business I couldn't get funded. I was living in a 900-square-foot studio in New York City with my college buddy, Karni, who was trying to break into the world of web development and moviemaking. I got the futon, Karni got the air mattress. It killed him that Flash, his chihuahua, liked to sleep with me more than him.

At the start of 2000, a company called Quo Vadis hired me to do business plans in Bulgaria, so I headed there.

If you like business, eastern Europe at that time was a blank canvas to an artist. My brain was on fire. That hustle mentality seeped from the pavement and the walls of the buildings the moment I landed in Sofia. So many things to be addressed and solved, not nearly enough people to do them. I met a man who was building a Voice over Internet Protocol (VoIP) telecom business for developing markets. At the time, no one in eastern Europe could make an international phone call without physical lines, the kind that go along the ocean floor. But VoIP required only a box to control "gateways," perhaps just sitting in a tiny Sofia apartment, and local phone lines to connect to the internet. The costs were internet use (less than a penny per call) plus a local charge (free). So a call from Bulgaria to the US, or vice versa, no longer cost 25 cents per minute, but less than 1 cent. Global telecom was exploding.

My network of developing countries soon expanded beyond eastern Europe to Africa, another part of the world that could really benefit from VoIP. Once my work was done in Kinshasa, capital of the Democratic Republic of the Congo, I headed to Tunisia but had to go by way of Brussels, where I missed the second leg of my trip. I was always living close to the bone, so I was upset that I now had to pay for a hotel. In the elevator to my room, the man beside the button panel looked familiar. "What floor?" he asked.

"Eight," I said.

"You sound like an American."

"Yes, I am. And you look like Richard Branson."

"Yes, I am."

"Holy shit!"

He was there for the launch of Virgin Brussels the next morning. He asked why I was there. When I told him I'd just come in from Kinshasa where I was doing telecom, he was hooked. We got out of the elevator and stood on the landing for a half hour talking about Africa. Virgin had numerous properties there and he loved it. Anyone with curiosity about the world would have wanted to know what it was like in Kinshasa then, but his interest went deeper. When we parted, he said I was a bright young man and did I have a card? I told him I would go to my room and get him my info immediately.

"Please leave it for me at the front desk," he instructed. Early next morning I wrote a letter on hotel stationery, my normally terrible handwriting so deliberately neat it bordered on serial killer. At the desk I handed the receptionist the letter and instructed, "Please give this to Sir Richard Branson."

The woman looked puzzled. "There's no one here by that name."

"Oh... no, I get it, but I'm, like, we met in the elevator last night and had a really good talk afterward, just the two of us, and he asked—"

Nothing doing. They wouldn't accept the letter. I didn't know if he was registered under the name Fred Flintstone or Johnny B. Goode, or maybe he'd already checked out.

I was crushed. I felt something special could have come from that. What a huge missed opportunity.

When I got back to the US, I FedExed the letter to his office. It was spring 2001. I wouldn't hear from him for 14 years.

The partnership with the VoIP businessman in Sofia broke up, and I was left with assets and no knowledge of how to keep it running. I flew back to New York, walked into a Barnes & Noble bookstore near my apartment, and bought *Deploying Cisco Voice over IP Solutions*, a 600-page tome that may not sound as exciting as *Harry Potter* but I devoured it, cover to cover, and learned how to program gateways so that I, myself, could turn them back on and run the network. There were investors to whom we—now just I—owed money who were getting antsy. I told them (a lesson that would spur me in future businesses), "I want to pay you, I'll work hard to pay you, but if you shut me down now, I'll never be able to pay you."

It seemed to work. They got their money back.

I merged the VoIP business with a phone-card company started by a college friend who lived in San Diego, so I moved there. I got set up on a date with Erin Lindsey, a production assistant at Fox Studios in LA. I drove up to Malibu for dinner with her and our mutual friends, Michelle and Jeremy, and that was it. Done. After a couple of more dates, I told Erin I was leaving for Romania for three months. I taught her how to text on her flip phone so we could stay in touch. When I returned, we split our time between San Diego and Los Angeles. I was more risk-taking than she was, and I hoped she thought that would make for a fun life together. I also hoped she would believe that I would eventually get my shit together. She was equally independent but more grounded. I didn't realize until I met her that my perfect companion would be someone who, if I had to go to Kazakhstan for two weeks, wouldn't blink, would continue to do her own thing, yet would always be there for me.

I don't know that my future father-in-law saw it that way. He was very corporate, a shopping-center developer, always digging for an answer from his daughter about what her boyfriend did. "What exactly is his job? Why is he traipsing around Azerbaijan?"

O

Everything needs fixing. Everyone needs fixing. You find things that need fixing when you travel halfway around the globe. You find them in your own basement. You especially find them when you listen carefully for the needs and wants that bind us. To communicate. To connect. To feel part of a community. To feel safe. To have peace of mind. The job is never done.

In October 2006, my dad was diagnosed with glioblastoma, a rare brain cancer with no known cure. I spent little time lamenting or just sitting with it: My instinct was to get entrepreneurial about it. There *had* to be a cure, or at least a better approach than what they were telling us. I combed every site and book for some way in. I sought out medical experts. The two-year survival rate for glioblastoma is not great—if you know someone with it, you may not want to Google it—and the one-year survival rate is pretty bad, too. Were there clinical trials going that my father could be part of? Could I help get a new one going? There had to be a solution.

Dad wanted none of it. He was a cautious man. He wanted to listen to conventional wisdom; I wanted to solve cancer. How was that rational? My father is dying and that's my first thought? How can I cure a disease that brilliant scientists and medical researchers have been working on for decades?

My father was getting weaker. It was the first time I truly felt the pain of distance, being on the West Coast, far from my family. My older brother, John, in Livingston, New Jersey, was all of 20 minutes from our folks. I flew out as often as I could, always trying to be practical—upbeat but forward-moving, proactive. But within just a couple of months of the diagnosis, Dad was already unable to walk unaided or drive. Anyone who knew him saw immediately that he was gravely ill.

That first Christmas after my father was diagnosed, I came to visit, and as soon as the cab pulled up to the family home, I was upset. Inside, I practically accosted my brother. "Where are the Christmas lights?"

It was unfair. John, by virtue of his proximity, had taken care of lots of problems that I could not from afar. But I knew my mother liked the lights; my father had grown up Jewish and claimed he was a "Christmas hostage" every year, but he not so secretly liked the lights and the tree too. He was the one who always put them up, with John's help.

I jumped in John's Ford Explorer, drove to town, and bought $500 worth of Christmas lights, ornaments, timers; the vehicle was stuffed to the ceiling with holiday merch. When I got back, I told John, "They can't be in this house at Christmas, just sitting there, thinking about his sickness." John and I lit up the frame of the house, as well as every nearby tree, bush, branch, and stray stick. By nightfall the place looked like one of those over-the-top homes that, when you plug everything in to light it up, it shorts out the whole street and everything goes dark. I stood there in the cold, seeing my breath, staring at our bright multicolor handiwork.

"Happy now?" asked John.

"Yes," I said.

I brought my father to Boston to see a brilliant neuro-oncologist named Dr. Santosh Kesari, then at Harvard's Dana-Farber Cancer Institute. But Dad wasn't up to it. He was probably too far along. In January of 2008, my dad, Bruce Siminoff, passed away. Erin was almost three months pregnant. Months later, I was asked to have dinner with a teen whose mom had worked with my dad. His father had also just died, and it came out that they didn't have the money to send him to college. I broke down at the table and told him I would pay for it, something I was not expecting to come out of me. When I got home, I told Erin we were paying for college for a kid I'd only met that night, with money we didn't have. But I had to do something with the pain of loss. I felt as if my dad would have been glad.

Later, too late for my father, I would seek out what was going on in brain-cancer research and what wasn't, to help support cancer studies and potential vaccines that were different from the chemo and radiation protocols and American Cancer Society research efforts that hadn't moved the needle in a long time, at least not with glioblastoma. The new studies probably helped save at least one life, hopefully many more, but not my dad's. I wish I could have saved him, at least long enough to see our newborn child.

The night before Erin gave birth via scheduled C-section, I attended a poker game hosted by Jason Calacanis, a tech-industry mover-shaker. It was a small group of LA tech people including then-local celebrity Elon Musk, the slightly pre-Iron Man version: The Tesla Roadster had just come out, and SpaceX's Falcon 1 was about to make its first successful launch. Erin wasn't mad at me for going. She got it. She knew that opportunity, unlike a scheduled C-section, doesn't work around your timetable.

And the next morning, beautiful Oliver Bruce Siminoff—Ollie, as he was almost immediately dubbed—was born. Right before Erin went in for the delivery, I joked with her about which amazing restaurants she wanted takeout from. Within hours after the birth, though, it was apparent that something wasn't right with Ollie. He didn't want to breastfeed, which of course upset Erin deeply. But the amazing nurses, who'd seen it all, kept cheering them both on. It was going to be fine, totally normal, happens all the time. Mom and newborn had a little success, but very little; the next day Ollie had jaundice. He stopped eating. The doctor began running tests.

Galactosemia, a rare genetic metabolic disorder. The newborn can't break down galactose, a simple sugar found in milk, both human and animal. If the infant drinks milk, galactose buildup can damage the baby's organs until they shut down and become infected with *E. coli*, and the child is essentially poisoned to death.

It was as if Ollie knew to stay away. The infant-mortality rate for newborns with the condition who drink breast milk is 70%. Mom unwittingly killing her child. Think about that for a moment.

Ollie was put on formula.

Galactosemia was an "orphan disease," rare enough that very little funding went to researching a treatment or cure.[7] The variant that Ollie was born with affects approximately 1 in 50,000 newborns, which means only a few thousand Americans have it out of more than 300 million.

Galactosemia was a curveball we were not prepared for. Would anyone be?

We went to see a genetic specialist. Our situation was uncommon enough that when he reached to his shelf of books on orphan diseases, he literally blew the dust off of one of them. "It's not just dairy he can't eat but legumes and vegetables, because those also have galactose," the doctor said. "Then there's the cross-contamination issue. Those born with galactosemia will have learning and speech disabilities. Some of them don't talk... "

My head was spinning. Erin's head was spinning. We were both crying. We had just had a baby, for Christ's sake... we had so many questions.

"Here you go," the specialist said, handing me the book, which was at least a generation old. When I was finally able to gather myself, I asked the doctor about stem cells. Vitamins. Treatments. What caused this? What was my son missing?

The doctor shrugged after each question, as if letting them roll off his shoulders. He was wearing a fishing vest. I had the gall to ask another question about my beautiful newborn with the orphan disease—

"What do you do?" the doctor asked, cutting me off.

"What?"

---

7   There are no drugs to treat galactosemia; there have been trials, but no approved drugs yet.

"What do you do? For a living."

"I'm... I'm an entrepreneur."

He nodded. "Great. Well, I'm a doctor."

I'm not sure where on the rage meter I was at that moment. He was lucky that my despair and sadness were so much greater. He all but said there was nothing to be done. There's *always* something that can be done!

Over the next days and weeks, the entrepreneur in me attacked the challenge before us, and I know I ticked off many doctors. They all told us nothing could be done.

I kicked that theory in the teeth.

It turned out that the formula Ollie had been put on was soy-based and contained galactose. We called everyone we knew for help until we got to Dr. Gerard Berry (whom we affectionately came to call Dr. Gerry Berry) at Boston Children's Hospital, maybe the only doctor in the world who specialized in galactosemia. He recommended an amino acid-based formula called Neocate that contained zero galactose. It wasn't easy to procure. A good friend from college, Jeremy Weiner, who seemed to know someone everywhere, recommended I call a pharmacist with a 100,000-square-foot warehouse in City of Industry, south of Los Angeles. I raced there and a security guard came out to meet me with cartons of Neocate. A true drug deal, just not the illicit type. Ollie seemed to do better.

We flew to Boston. Compared to Fishing Vest Asshole, Dr. Berry was like night and day—his willingness to consider options, his compassion. My questions yielded answers like "That might work... hmm, we never thought about that..." Did I consider this top doctor less than well-informed because he was open to new plans of attack for a disease that, frankly, *no* one knew much about?

No. I considered him a scientist. And a human being. Someone trying to look for new rules because the existing ones weren't working. It's what you do as an inventor.

The doctor suggested a brain MRI to get a better idea of what was going on with Ollie; it might show lesions or other abnormalities. And even if we found those, the prognosis did not have to be dire.

"Well, let's get an MRI!" I said, pumped.

The only place that did brain MRIs on such young subjects was Children's Hospital Los Angeles. The procedure would be guided by Dr. Berry.

Ollie's MRI was encouraging: Yes, he had galactosemia, but it was nothing like a death sentence. The doctor's first recommendation after seeing the MRI results? "He should be fine. Don't give him a glass of milk. Maybe he shouldn't eat certain other foods, but if he has a little bit, it won't kill him."

That gave us an opening to live our lives. When Ollie was a little over a year old, we went to a speech therapist to work with Ollie prophylactically. She said we were too early and to come back in a year. Instead, we found another one who would take him on. Like so many parents, we were willing to do whatever it took.

All Ollie needed to avoid was milk; everything else, including eggs, seemed fine. When we met parents with kids with the same condition, they were shocked to hear about our day-to-day. "Wait, you go to *restaurants*?" I didn't have the heart to say that we ate out several nights a week.

We realized how lucky we were that Ollie was doing as well as he did and didn't manifest many of the more serious symptoms that we had been warned about. Maybe it was the Neocate. Maybe it had nothing to do with that. So much was unknown, so we attempted to learn everything possible and follow the best plan we could identify. With the support of Boston Children's, and the entrepreneur in me—and Erin—we uncovered all kinds of helpful resources you might not know about if you didn't push open the door behind the door behind the door.

Then again, did it help all that much? Erin and I always wonder if what we found made the difference; we always just thought of Ollie as our miracle child. That's what the incredible people at Boston Children's call him when we go for an annual checkup. In his most recent semester at school, Ollie made the honor roll. There's never been a greater kid.

Every time Ollie accomplishes something terrific (makes the dean's list, gets invited out of state to play in a basketball tournament), I think about that arrogant loser in the fishing vest who talked down to Erin and me and told us there was really nothing to be done. Those accomplishments of Ollie's are the greatest "fuck you" ever.

But I don't focus on that. I focus on how grateful I am that my favorite human on the planet is thriving. How the intensity with which I work on and fight for my ideas and inventions is nothing compared to how Erin and I fought for him and will always fight for him.

I love fixing things. Ollie fixed me.

CHAPTER 8

TOO DUMB TO FAIL

One thing had become clear to me, more than any other. We at the company were a team. Close-knit, often yelling at one another (okay, mostly me doing the yelling), always there for each other. We spent an insane amount of time together, had already experienced so many ups and downs. We had been through stressful times and great fun, and worked hard when we worked. I'd always hated office cultures that demanded you be on the job or in the workplace a certain set number of hours just because it *looks* productive, or because that's how the people before you did it. Fuck that. Did you do the job or not? Did you deliver what the customer expected? That's all anyone on either side of the transaction should care about.

I'd also never been one to throw parties for every little milestone, because that signaled the job was done. And the job is never done.

To celebrate our successful rebuild at the start of October 2014, I emailed our original customers.

> Hey, you bought a DoorBot. We did our best on the product, but we understand it had its difficulties. Here's a promo code for $100 toward the purchase of a Ring [$199].

"Did our best"? Yes, that was true. But did it pay to be that honest about a deeply flawed product?

Some people thought it was reckless of me to continue to have my email address, j@ring.com, on every box. But I wanted there to be no one between me and any customer who wrote to me. Why is that the exception to the rule? Why doesn't every CEO and founder do that? If you're too busy to communicate directly with your customers, then you've cut yourself off from your single most useful source of information. That's how you truly know if what you're doing is right or wrong. And having a team of people to always "field" or "filter" the communications is just sad. It's the difference between ground truth and watered-down truth. If you haven't heard your customers say what they're actually thinking in their own voices, if you don't make it incredibly easy for them to tell you, then you should probably find a different line of work. I take so much pride in that. I consider it the greatest hack for any person leading a company that has a public.

Though it shouldn't be called a hack. My way better-resourced competitors often made things easy for us. For example, some of them created the most gorgeous sets at the tech conferences we all attended, and the most beautiful products, an aesthetic polished inside the bubble of Silicon Valley. They were exquisite... and because of the cost to make them, totally unattainable for the majority of average Americans. To make great products you need to empathize with customers. That's your first and last job, but it's a tough one. People think only in their own way. To think in someone else's way takes effort. Time for some of those Silicon Valley product designers to fly coach once in a while, on a flight that lands in the middle of the country, not just the edges.

My inbox flooded with responses to my $100 promo offer, a mix that ranged from excited:

This is awesome! Thank you!!

to excited and looking for a better deal:

Hey, if I buy 3, do I get a 4th free?

to brutally honest:

> I am glad to hear I helped finance your engineers and new product. Due to issues with Doorbot, I don't think I am interested in going an additional $100 in the hole. I can understand why you changed your company name.

I wanted everyone to feel heard, whether they were psyched about the offer or hated me and needed to vent. For the latter, depending how epic their contempt, we sometimes offered a promo for the new version, free. There were lots of DoorBots out there performing less than perfectly. I was eager to take as many out of circulation as possible.

Then again, with the switch between our old, flawed product and our new, far superior product, we were overwhelmed. We were getting so many emails, and extraordinary times call for extraordinary measures. I took the rare, uncomfortable step of having August and Dave log into my email account and respond as me, at least to some of the messages.

Every now and then, August or Dave would mistakenly sign the email, "Thanks, August" or "Thanks, Dave" instead of "Thanks, Jamie" and hit send before they caught their mistake.

"Damn," said Dave, after sending another email with his signoff. "Did it again."

Served me right. So much for the founder's personal touch.

Fact is, I tolerated all kinds of mistakes from my team. Yes, I held them to a high bar, and I had a tendency to fire people, but it was for doing something they needed to learn not to do, and when I rehired them (almost always within 6 to 24 hours), we both knew they would never do that thing again. If I didn't fire them, I would definitely let them know they had screwed up; I wouldn't pretend it was just a glitch. Ours was boot camp, a great place to learn how to be good and tough, both. Or, to use another military analogy, they each came in as a piece of steel and over time became stronger, sharper, more honed.

Working at Ring was Darwinian. The hard, resourceful workers had unlimited upside. The ones who didn't fit just didn't fit. That didn't make them bad people or even bad workers. It simply meant that our environment wasn't for them. An all-encompassing work culture might sound like a beautiful idea, but really, how weird would it be to have an office where every type of personality had a place?

Many of my best people I'd chosen precisely *because* of their youth and inexperience. Without that combination, we wouldn't be standing. When hiring, so many bosses look for experience. I looked for the opposite. For me, the greener, the better. Kids just out of college didn't always do the obvious thing when confronting a problem; sometimes they took a path that was more interesting and ultimately effective. A college dropout solved our supply-chain issues. A design major cracked the engineering challenges for almost everything we built. A recent college grad became a public-relations master in just months. Youth and inexperience were among their greatest assets, not liabilities. I didn't stew over their mistakes because the best people held themselves accountable. They pre-punished themselves and figured out how to be better before I had the time to rip them a new one. You're a young, inexperienced worker? You've got it made. Here's your chance to fail, show how quickly you realize that, and go and correct it.

Not that you had to *be* young. You had to *think* young. Some people just do. Open. Ready for anything. That's the part that mattered.

Smarts mattered too, of course.

What mattered above all was passion for the mission, the single most important "qualification" when I interviewed someone for a job. Do you have energy for the mission? If you do, you'll figure out all the other stuff *because you care. Because you absolutely can't not be there doing what you're doing.*

If that's the case: Come work for us. We love you.

Another reason I valued youth and lack of experience over experience: the willingness to be thrown into different jobs, constantly, and figure it out. Suppose I hire you to do PR when we're a $3 million company. What happens if next year we're a $30 million company—can you do PR for *that* business? It probably requires a bunch of new skills. What about when we're a $300 million company?

I needed people who could adjust in real time, because Ring was like a snake shedding its skin. We were growing so fast that we were essentially a new company every six months. Could you adapt, again and again? Some did. Yassi did. Dave did. August, John, and Karni did. Some didn't.

Most people believe that hiring more experienced candidates in tech, where things move so fast, enables you to get up and running sooner. You save on training time. But in most cases I found that to be wrong. A disruptive company needs to train *everyone*, or everyone needs to train themselves, because everyone is doing something they've never done before. That includes engineers. Sure, I could have limited our camera technicians to only those with impressive, relevant accomplishments (a history working at GoPro, for example). But I didn't have the money to pay those people what they could demand. Experience can be great. But experience also means the person has seen how things worked in the past, seen the steps to a solution. The world changes so rapidly, and we were trying to innovate and even pioneer, so a "Yep, seen this before, I got it" mindset and type of solution might be obsolete. On the rare occasions when I hired those with experience, if they *did* screw up, I fired them faster.

(I eventually did fire the engineer who had caused the Christmas Eve-Eve disaster that messed up the doorbell video and nearly took the company down. But it wasn't for that mistake, or not *just* for that.

Everyone makes mistakes. He was sincerely trying to make things better. But when you keep making mistakes, then maybe you can't call them "mistakes" anymore. I wished him well.)

As a leader (I hate the word "boss," I hate the word "employee," I hate "all-hands meetings" and "executive meetings"), there was nothing I liked more than finding diamonds in the rough. They're out there, all over the place, across all disciplines, and they're there for anyone to pluck, if you know what you're looking for, know how to nurture talent, and approach things with an open heart. Tom Brady, winner of seven Super Bowls and the greatest quarterback of all time, was chosen 199th in the 2000 NFL draft. Anybody could have had him. Every single team in the NFL passed on him, not once but several times, including the New England Patriots, the team that ultimately picked him. Even the Pats initially saw Brady as their third-string QB, at best. According to "expert" talent evaluators, he was a long shot, an afterthought, a nobody until 198 other players were chosen, most of whose names no one but their families and close friends now remember.

To be fair to the experts, even though Brady went to a big football school (the University of Michigan), he was not yet what he would become. Still, what he would become was there all the time. It just took someone to see it and believe in it. And then Tom Brady became Tom Brady and changed the history and perception of the Patriots. And football.

What did I learn from Tom Brady going 199th? Don't shoot too high. You don't have to compete with everyone for the consensus best person. You just have to be able to spot and foster talent. Find young people with promise, a spark of some kind. Look for the ones who might get overlooked, for whatever dumb reason. They didn't graduate from an Ivy League school. Maybe they didn't graduate from any school, period. They don't have tons of degrees. The Ring team was no one's idea of a collection of first-round draft choices. You build things with Legos and you're passionate? You're hired as my designer. You're sleeping on someone's couch but you're good with people? Sales. You went to a scrappy business school and did everyone's homework to make beer money? Name your job, I don't care, I just want you on my team. You've never been a CMO before but you understand story and brand? Why not, let's give it a try. My hires turned out to be no less qualified to do what we were aiming to do than a bunch of Stanford supergrads. We were just as capable of kicking everyone's ass, as long as we outworked the competition, we got a little bit of luck, and I could always find us some more money. There are lots of Tom Bradys out there, and if you're a Tom Brady at picking talent, you'll find them. At so many Silicon Valley firms, people are touted for all the different successful companies they've previously been associated with. "Sophia was at A and then B and then C and then D and then E, and we were so lucky to grab her when they offered her a job at F... " Yeah, Sophia has done some things. I want to find Sophia before she gets to A, not necessarily when she's with her 14th company.

None of us at Ring were above doing any task. Soon after he joined, Spiro Sacre, my eventual brilliant Head of Product, came to me about one issue but secretly lobbying for another. We were overflowing with garbage because our cleaning service came only once a week, which was fine when we were 12 people but inadequate once we'd grown to 45. "You think we can figure out what to do about the trash cans overflowing?" he wondered.

"I do!" I told him, echoing the approach I took when Dave told me we were low on cardboard boxes. "Thank you so much for volunteering to handle it!"

(Soon after Spiro took out the garbage a couple of times, even I had to admit it was a problem and broke down and had the cleaning service come twice a week.)

True, some skills we simply lacked because of my hiring tendencies. But it also meant that when a team member figured out that skill (forecasting inventory, for example), they might do it in a creative way that hadn't been thought of before. One of our youngest team members once said, "We're too dumb to fail"... and every single one of us took it as a compliment. I should have had T-shirts made. Since we weren't experienced enough to realize how bad things could end up, how potentially disastrous some of my choices were, we just kept moving forward until we figured out a solution. Fake it till you make it.

The principles of our culture were just word of mouth, at first. But eventually we were hiring so fast that I had to "go corporate" and actually write them up for all the new people to understand who we were:

- **We're DOERS.** Doers are "first principle" people. They just get shit done.

- **We are all adults and we are all equal.** I hate the idea of levels. We do not have titles. If you want one, fine, but be prepared to do anything.

- **Make neighbors happy**. Our customers are "neighbors," our most important contingent.

- **NO POLITICS!!!!!!** We have no room for office politics or BS.

- **We are here to win.** This seems pretty self-explanatory. It better be.

- **We care about each other**. We can fight and argue, and we do all the time, but there's no need for meanness.

- **Believe in the mission**: Reduce crime in neighborhoods. If you don't believe in the mission, that's fine. You're just not right to work here.

- **We don't celebrate the victories; instead, focus on the issues.** The job is never done until 100% of crime is solved.

Our work culture wasn't for everyone. It was fun but very hard, messy. Due diligence and following code were not among the top priorities. If you were buttoned-up, we probably weren't for you. You had to work in a building painted Ring orange. We were inclusive, diverse, weird, different.

Those that loved the culture *really* loved it. They all became antibodies, searching for the foreign objects trying to hide among us. The believers pushed them out. If someone wanted to play politics, they didn't stand a chance. Bye-bye.

If you took things personally, we were not a great fit, especially if you spent time around me. I had trouble relaxing. I needed to be more chill. Things at Ring ran hot and cold, rarely warm and cool; someone described me as a light switch without a dimmer. I put multiple holes in walls (covered with Post-its) and cracked a few smartphone screens. I was known to be thin-skinned if an outsider wrote something about us that was untrue, but don't you want your entrepreneur to be wired like that? We had no human-resources department and wouldn't if I could help it, and not because they would have investigated the time the Ring founder threw a phone at the head of one of his people. (I may not have been skilled enough to play high school baseball, but my aim is good enough to miss someone if I want to.)

I won't let myself off the hook for my "moments," but I felt stressed pretty much every minute of every day about all the money we were spending, all the investors I owed, all the people I'd hired whose livelihoods I felt responsible for, and above all the neighbors we were there to try and protect better. A good night's sleep was a foreign concept to me. Trail running in the woods behind my home was the closest thing I could find to relaxing. To be honest, the only way I might truly be able to let go is if someone fired *me*. I would be released. But I held myself to the highest standard, I tortured myself, I couldn't sleep and couldn't stop—and I also lacked the ability to leave. My life was all in on Ring. During the hardest days, I sometimes felt it was my misfortune that getting fired was not an option.

I had tendencies: I was more likely to ask for forgiveness after the fact than permission before. (And I forgave others quickly and often, a necessity given that I fired almost everyone in the office at least once, especially those who'd been there longest. August claims he holds the record, 10. My own guess is that John beat that easily but was afraid to count them up.)

I never minded spending money we didn't have, if it was for a long-term good cause. That's the difference between spending money and investing. If we failed, I was doing it in a spectacular way, not just petering out with no one noticing. I was not always aligned with VCs, but on this one we were in lockstep. They didn't want a small safe return but big bangs.

"Work-life balance" was a phrase and a goal, not a reality, though I considered many people with whom I worked to be friends, too, and the transition between "at the office" and "after work" felt fluid to me. I took Ollie with me on work trips to China and Europe. Erin was not happy that I hired some of Ollie's school friends' dads. What if everything blew up? A failed company was one thing, but possibly damaged relationships with the families of our son's friends? How bad would that be?

A lot of times I just went with my gut. I was wrong quite a few times—like that electrical engineer hired through a recruiter who messed up the video for thousands of DoorBots. Or the CTO who kept quitting. Or our "clean room." But I would rather be wrong sometimes with quick gut decisions than stuck in indecision. It was almost impossible to manage the number of decisions we had to process each day. I gave massive autonomy to my leaders. We were 40 CEOs—excuse me; *leaders*—rowing in the same direction, to fulfill the mission. At Ring we did our own thing and most of the time we got it right.

One reason our environment worked is because I had something to prove, as did so many of the people I hired. (See also *Brady, Tom.*) Was it because kids in middle school made fun of me? Because I spent a lot of time alone inventing things? Who knows?

If you liked classic, intense startup culture, that idea of keeping the train moving while also laying track down just a few feet ahead, then you loved working at Ring. We shared a deep camaraderie and loyalty. When someone showed me that an anonymous ex-Ringer had left a critical comment on Glassdoor, a popular job forum, describing in detail how tough it was to work at Ring, I shook my head. "Totally wrong," I said. "It's *way* harder than they're making it out to be."

I paid my people decently for a startup (meaning terribly for a non-startup). I kept my own salary artificially low, which turned out to be a mistake. At first it was a point of pride ("I'm the lowest paid exec here!" I often boasted), but when it led to financial worries for me and Erin, it seemed a dumb hill to die on. Still, if all of us accomplished what we wanted, then the equity one dreams of with startups would be there. I hoped.

Maybe best of all for making a great work culture: Everyone on my team knew we weren't selling "just" PCB boards, plastic, or screens. We were selling a sense of safety and peace. We were selling a dream of better neighborhoods and communities. It shouldn't have been a surprise that the video doorbell, the most promising idea to come out of my garage, was born of need. We were driven by a mission. We didn't have to force it. It was real.

And hard as the work was, it wasn't as if there were no moments of triumph while we toiled. If you joined in the months after my *Shark Tank* appearance, you got a taste of success fairly quickly. You could see our product on store shelves. People talked about it. We were an ideal home for a ragtag crew of dreamers and scrappers, underdogs and misfits, who didn't seem to know, at least at first, what we were doing.

Yeah, we needed help.

O

We did more than $600,000 in orders the first two weeks after announcing Ring (4,300 units, retailing at ~$199), then slowed to around $14k per day. Our first production run of 5,000 units from Foxconn was almost sold out. We could ship 8,000 more doorbells before Christmas and had components for another 20,000.

We had spent no money on marketing, and the media coverage was phenomenal. (I finally let Yassi send out product to reviewers.) There were still some embittered DoorBot customers out there, but for the most part we were getting great press. Why? Because we were underdogs. People wanted us to win.

Our new product was really good. And it answered a need.

O

At the (rare) office launch party for the Ring Video Doorbell (no alcohol, daytime, have to get back to the mission), I tried recruiting my friend Josh Roth to come work for us. I couldn't deal with my CTO quitting on me over and over (or threatening to quit on me). Josh nodded in understanding. He was being recruited by another company, but he came to the Ring office, saw our operation, and met with some engineers and our finance person to see that our money situation was viable, at least for the next 48 hours. He was impressed with who was buying our product— good demographics. At the party, he walked outside with August, whom he knew from Unsubscribe.

"I can't believe you're here," August told Josh (I learned later). "I never thought in a million years you'd consider working with Jamie again."

"Tell me something," Josh said. "Which Jamie is this? The Jamie I couldn't stand and bothered me so much at Unsubscribe because he was working on twenty-five things? Or the Jamie who's ultra-focused about what he's doing?"

"I've never seen him so focused because I've never seen him so driven," said August. "It's all he cares about. He's working on something that matters. We all are."

Josh nodded and looked around the space. "This office is just a bigger garage than he has at home," he said, and left. August was unsure if the comment was positive or negative.

Josh began consulting with us that month, overseeing the hardware team and firmware team, helping with logistics, supply chain, and finance. The engineers were particularly happy he was coming on, not only because he was good at what he did, but because they knew he would probably act as a buffer, cushioning them from some of my "intensity" when I got mad. Josh had been a successful CTO when we first met on the beach years before. He'd been computer science and engineering out of UCLA and had worked at a number of places before he and I partnered on Unsubscribe. He had a young family and was better at focusing than I was. He would start full-time for us after the holidays.

And then I really needed someone who could sell anything to anyone. As much as we'd pushed our sales online, we needed to get our product beyond Ring.com and Amazon and into the big retailers like Home Depot and Best Buy, Target and Walmart. Don Hicks was my kind of guy. He'd left college to join the Marines, fallen off a 50-foot tower in boot camp and hurt his back, gone to night school, sold Nextel telephones door-to-door until that business collapsed, sold soldering irons to OEM manufacturers and then to big retailers like Circuit City and Sears, sold board games and DVD games to Walmart and Target until *that* business collapsed (thanks to Apple opening the App Store), then consulted on other businesses before a mutual friend told him that he should talk to me because I was desperate for an *experienced* head of sales. (Yes, sometimes experience is a huge plus.) His philosophy was simple: "Go sell shit." We joked about the possibility of selling doorbells door-to-door... do you ring the doorbell? If they have a working doorbell you want them to replace, do you knock?

Don had been through ups and downs and always figured out the next thing. He wore loud clothes and a shiny gold watch three sizes too big for his wrist. I hired him immediately, of course.

The year had been a mix of highs and lows, something I could truthfully say about each and every month we'd been in business. We had done just under $1.3 million in sales at our website. The only brick-and-mortar outlet with the product was Brookstone, which had sold 500 units. I told my team that I hoped to get to $8 million a month within a year, an insane jump from our current level. Then again, Don "Go Sell Shit" Hicks had jumped right in and already won commitments from the big-box stores to take the new product by March, and projected us to be in 6,000 stores by the end of the year. In February, we would pitch again on Home Shopping Network. The buyer there tried our product and upped the order from 500 units to 2,000, and scheduled their top host to do the first segments. It looked like I was in serious jeopardy of losing my Land Rover to Diego.

Time magazine came out with its year-end "Top 10 Gadgets of 2014." So grateful they don't do "Top 9" lists. Our video doorbell slipped in at #10.

The next planned product was a plug-in chime that many of our customers had been asking for. Josh was pushing it. I wasn't sure if we were ready to do an accessory. Would it distract us too much? What the hell, we went ahead with it. You get zero hits if you take zero swings, so we kept swinging.

A couple of weeks before the holidays, my cousin Mark and I climbed onto the roof of the Ring building to string Christmas lights, neatly avoiding the trash bags duct-taped down to stop leaks. As I draped another strand of multicolored lights, I looked at Mark. "If our Jewish grandmother could see us now," I said.

All in all, I rated the Ring Video Doorbell, codename F5, the new product we'd been working on for almost a year, an F4.5. The hardware was F5, but the software had some bugs, more like F3.5. Of the 8,000 new doorbells we'd sold, 3,000 had come online, and about 10% of those had experienced some issue. We'd made sure to test all the units before shipping them out (a big difference from a year earlier) and regulated the rollout of new units to make sure our quality-assurance protocol was impeccable. I gave us all two months to get the product to an unmistakable F5.

Not bad for a bunch of misfits who were too dumb to fail.

CHAPTER 9

MORE THAN A DOORBELL

The Consumer Electronics Show in January 2015 was so different from a year earlier. People knew who we were. They checked us out, probably to copy us and eventually to outright steal.

I know that sounds arrogant. But we were right about F5. We *had* to launch in the fall of 2014, front-running CES 2015. If we hadn't come to the show already in the market and leading it, ours would have been just a "me-too" product, not *the* doorbell, the one against which all the others were being judged. None of them surpassed or equaled us in technical quality or look. But what scared me was the sheer number of video doorbell copycats at the show, somewhere between 30 and 40. I panicked that we might be overrun simply by the mass of competition, shoddy as their offerings were. The flood briefly terrified me.

And then it didn't. Being first to market, then hammering away and not giving up, was huge. For the journalists covering our sector of the tech industry, it was Ring and everyone else. Not just for coverage, but awards, too.

What made our job easier: None of those look-alike smart-home doorbells understood what business we were actually in. What truly separated us was not our quality or our features, great as those now were. It was our mission, what we stood for. We were not just a doorbell company but the company whose technology made you feel safer. *Brand, brand, brand.* Customers looking for a video doorbell in the aisles of Home Depot don't know or care about the difference between 720p, 1080p, and a chocolate donut. We needed to give them a simple reason to seek us out above all others.

And our booth this time was a real one, not just flattened boxes for walls. Dave and I built it right there in the convention hall. We had turf, naturally, plus an actual tree and a water element we built ourselves. A no-necked Teamster came by to tell us that if we didn't stop, the screwdriver I held might be going through my head.

Dave and I kept building. Empty threats. No screwdrivers shoved into any heads. I was more scared of failure than union violence. I was scared of mediocrity and middle-of-the-packness. Also, personal bankruptcy.

But no one was getting in the way of what we were trying to accomplish.

I couldn't help myself. When Dave and I were done building our booth, I walked over to yell at one of the more egregious of the doorbell copycats. I already knew about these particular thieves, because they had copied our old GetDoorBot.com website. When I say "copied," I mean virtually word for word; their site was quite literally ours, except they had done a find-and-replace and put their name wherever ours had been. The people in the booth stared at me while I cursed them out. They weren't native English speakers, but they got the point.

To play our demo videos, we could have rented overpriced monitors and sound systems from the convention folks, the way most vendors did, but as I had told my team over and over, every dollar we spent was really a hundred, if we hoped to deliver a 100x return to our investors. I told Yassi to go to the nearest Apple Store on the Strip, buy five of their biggest Macs, and bring them to the convention hall. We'd set up the Macs, play our demos, and then pack up the computers for her to return them to the store the next day.

"Are you serious?" she asked.

"Totally. Saves money. I've done it before. You get really good at repacking. You just tell them it wasn't what you had in mind." I'd actually been doing it for years. It was the best solution. On one hand, nobody wanted to lug a giant monitor to the show; on the other, the ones rentable from the venue tended to be beaten up and overpriced. Why do that when you could buy five shiny new Macs for a day? "And don't look at me like that," I told Yassi, though she had every right to look at me like that. "Apple has a built-in restocking fee of five percent, so it's more like a rental. We're not stiffing them. If they didn't build in the fee, I wouldn't do it." I considered. "*Maybe* I wouldn't do it."

Retailers flocked to our booth. We had lots of visitors from the smart-home market, all those touting "the Internet of Things." Traffic was great. No longer were we relegated to a dark corner. No longer were we reduced to blocking people's paths as they walked past us, making them look our way while we pitched them at 4x speed.

Once, when Yassi was presenting, the sleeve of her shirt lifted, revealing her arm tattoo, and she quickly pulled the sleeve down as if embarrassed.

"You don't have to do that, you know," I told her later. "Be you. That works best."

In fact, that was the perfect metaphor for us, I thought. A tattooed arm *was* who we were.

I'd waited until the last minute to rent rooms in Vegas so all I could find for the team was an Airbnb in a suburban development about 15 minutes from the convention center, a house owned by the self-proclaimed "#1 gigolo in Las Vegas," who turned out to be a charming guy who had different types of cars for the different women he escorted (Jeep for one, Mercedes for another, and so on). All I could find on the Strip for me and Don, my head of sales, was a single room with two queen beds at the Wynn. We needed to be close to the action because we had meeting after meeting, deep into the night. I did my thing, but I marveled at Don, who was an absolute bulldozer, the way you need your head of sales to be.

CES 2015 was such a success for us that I superstitiously attributed an unusual reason for it: It happened because Don and I shared a room. I told Don that from now on, he and I were sharing a room whenever we traveled together.

We would do that for the next four years, in hotels around the world.

O

By mid-February we'd sold 11,000 units, about 8,500 of which were "active," meaning operative and online. Our return rate was low (2.5%), a clear and encouraging trend since we'd started shipping the new product. The glitches were now random anomalies, which we fixed as we found them. We would get the return rate down even more with the next firmware update.

I wanted to change the packaging for our spring push into retailers. And we were working on two new products: the wireless chime and a wired-only version of the doorbell.

Maybe most exciting, we launched free "cloud recording" to rave reviews, which yielded useful and revealing—and hypnotically rewatchable—video content, like the USPS worker tossing packages over a customer's fence. Simon Cassels, father of one of Ollie's pre-school friends whom I'd met around the morning singing circle, was a creative director at Apple with an innate understanding of story, and he called the Ring doorbell "a weapon of mass communication."

"It records the story, people share the story," he said, impressed. "You have every form of marketing and advertising built into your product. Every time you sell a doorbell, you have an outdoor billboard. Normally you have to pay for a billboard. You put one on every single door of every home, and it's a conversation starter, an experience."

I nodded in agreement. Then I tried to hire Simon to be our chief marketing officer. He declined, for the time being; he wasn't available (plus I couldn't pay him what he would get on the open market). But, like my cousin Mark, he was eager and willing to help us when he had the time. Like me, Simon got the obsession with Dyson, and how they were almost the perfect combination of high-quality, innovative products (vacuum cleaner, in their case) paired with a name, reputation, and maniacal focus on the consumer that set it apart from all its competitors. We talked about the need to be the Dyson of home security: Ten years down the road, when a consumer walked into a Best Buy looking for any home security–related product and there were 10,000 competitors all screaming for attention, Ring needed to stand out. When the consumer saw the Ring name and Ring packaging, their thought had to be some version of, *That's going to make my house safer. Me. My family. My neighborhood.*

In short: We needed everyone to channel Erin's first thought when she saw the DoorBot prototype.

If within a few years the name Ring didn't trigger that kind of reaction, we'd be dead. And deserve to be.

We would start charging soon for our cloud-recording feature, which had been a free trial for the first few months. I thought $5 a month sounded steep. The idea was that greater safety should be available to everyone, not just those for whom an extra $5 a month means nothing. We had no idea what our "attach rate" would be—the percentage of consumers who would pay the monthly fee to have their content saved for six months—but the success and future of a hardware business like ours more or less hinged on it. The difference between, say, a 10% attach rate and a 30% rate was obviously massive. Existential. If we hit a threshold, then our business would survive and potentially thrive. If we didn't? There was always bartending.

One morning, Yassi came in excited. Her parents' friends in nearby Encino had a Ring and they had stopped a burglar in his tracks thanks to their video doorbell.

"Holy shit!" I said. This could really work!

O

We had the battery doorbell (retail $199). We were about to come out with our first wired doorbell (code name: "The $99 Doorbell"). We were inclined to price it cheaper because it was less expensive to manufacture. Yet it was sleeker and more handsome, so it looked as if it should cost more.

That was a big problem.

Though I don't love meetings, I gathered a small group around my surfboard table to have it out over price. We set the two doorbell models next to each other. We were getting nowhere until the food-delivery guy brought lunch.

"What do you think about this?" I asked him, pointing to the two units. "You see these two products in a store, what's your first reaction? Which one are you picking up first?"

He went right for the better-looking, wired-only unit. Which cost far less to make.

I thanked him and gave him a big tip.

It didn't feel right to price the better-looking one at half the price of our main doorbell. Yes, the sleeker one could be used only on houses wired for doorbells (about half of US homes), and it required more installation. But if we really started selling lots of the wired-only ones, that might deeply cut into the battery-powered doorbell, our bread and butter. If there was one business lesson I knew growing up, learned more formally at Babson, and learned again and again as an entrepreneur, it was this: *DO NOT MESS with the thing that people have already shown you they want to buy.* A hundred years after being the most beloved brand on the planet, Coca-Cola decided in 1985 to switch to New Coke. Less than three months later they realized what a dumb move it was and switched back to the formula that everyone had loved for the past century.

We couldn't have people suddenly stop buying our flagship product. So how could we solve the pricing problem? Erin, Ollie, and I were returning one Sunday evening from a ski weekend at Mammoth Mountain, and I almost drove off the road when the answer came to me: *Of course! So simple!*

What's another basic business lesson you learn, about branding? *Perception is everything.*

I called everyone who had been in the meeting with the food-delivery guy. "He thought he was taking the better one, the one we're planning to price way cheaper than the other one," I said. "We can't do that. This is not about pricing based on what it costs to make. It's pricing based on perception. We have to give people what they *think* they're getting."

In business, perception is reality. Fighting it is like fighting gravity: *You will lose.* We had to work with that unchangeable truth. It makes total sense to price your products based on your cost to produce them, but in this case, we had to raise the price of the wired doorbell to be more expensive, because it *looked* more expensive. We packed a few more features on it to legitimize the price, and named it the Ring Video Doorbell Pro.

The wired-only version—with the mislabeled codename "The $99 Doorbell"—would retail for $249, with the battery doorbell sticking at $199.

O

Like many 6-year-old soccer players, Oliver "Ollie" Siminoff spent more time sitting in a corner of the field picking blades of grass and trying to catch bugs than running for the ball, or even looking up when it whizzed by his nose, despite his father and mother yelling to him that he might want to join his teammates at the other end of the field.

Which was fine with me, as Erin and I sat on the sideline one Saturday next to my friend Trevor Bezdek and his wife, Jana, whose son was also out on the field, nominally playing U-7 soccer.

Not surprisingly, talk turned to work. Trevor had co-founded the fast-growing drug coupon site GoodRx along with Scott Marlette and Doug Hirsch (like Scott, an early Facebook executive). He was explaining the company's latest jump into the stratosphere. "We started doing TV and we're crushing it," he said, almost in disbelief. He told me just how much his numbers had been turbocharged. Searches. Sales. New customers.

I felt like a man in the desert who comes upon an oasis. Was there that much difference between drugstore coupons and wifi doorbells?

I called out to Ollie, who looked up from his nature project, and blew him a kiss. I told Erin I had to go. I raced to the office.

We had to do TV.

Over the next few days, I discovered that not everyone in my circle agreed. They thought commercials were a wasted expense.

"You're too small."

"It's just not done."

"Stick to Facebook, maybe pump a little more into Google AdWords, like everyone else? Digital's so cheap. You are not ready for TV."

"Until you get to a hundred million in sales, you CANNOT be on TV," someone warned me, as if it were an actual rule. "You'd just be throwing away money."

I hate tautological BS, that "you just don't do something because it just isn't done." I thought of the fights Erin had to win at her studio to get a movie made of John Green's novel *The Fault in Our Stars*. Never mind that the source material was a huge bestseller. She was warned, "You're gonna kill a fucking teenager *and* she wears an oxygen tube the whole movie? Are you kidding? No one's watching that fucking movie!"

Yeah, well, with a $12 million budget, the movie earned $300 million, one of the highest returns on investment in the recent history of Hollywood.

My instincts made me want to spend the money on TV. The pushback made me want to even more.

I was "advised" by another expert that if I was serious about a slickly done TV commercial, it would cost $800,000 just to produce it.

Yeah, no. We would find another way. And I had to find it with limited resources. I was inspired by the example of the guys at Dollar Shave Club, who used clever, inexpensive marketing techniques to get penetration.

I once again recruited Karni, who'd already made great videos for DoorBot. One week and $5,000 later, we had a 30-second commercial. We bought time on HGTV and DIY Network. We measured its effect.

Crickets.

We spent tens of thousands of dollars on TV those first months. August tracked our progress. The commercials seemed to have a minimal impact on sales.

We had some money in the bank; we were selling decently online; Foxconn was still nice enough to let us take 120 days to pay them. But the fact is we owed millions. My spending—all worthwhile investment, to my mind—was eating cash faster than the growth could cover. Red was closing in again. One of my VCs wondered out loud, "What are you doing? You're a tiny little company doing TV commercials?" I'm sure others were thinking it, too.

True, measuring ROI for TV commercials could be tough. There's a famous saying in the advertising world: "Half the money we spend is wasted but the problem is, we don't know which half." Could the money I was spending on commercials be used more effectively elsewhere? Maybe. I hoped enough people would start seeing enough of our doorbells on other people's homes to check us out. I hoped satisfied Ring neighbors were telling their un-Ringed friends how great our product was. I knew the Ring videos all over YouTube helped.

And then, two months after running our first commercial, we noticed sales start to tick up, in direct relation to the commercials: Soon after every airing, we saw a sales spike.

Holy shit, I thought. *It works! I love direct-response marketing!*

I told Karni to make more $5,000 commercials. He would produce them, Simon would be the creative lead, I would be the spokesperson. Once again, my house and backyard served as the set. Lots of Ring staff helped. I recruited friends and neighbors as extras. One of our early commercials was a reverse scare tactic: Night, three guys dressed in black, two of them masked, one of them breaks a window at the back of the house. As they skulk through the house with flashlights, I'm sitting on the couch, casually pointing to the intruders. "They want you to think this is what a home burglary looks like," I say. The next shot is daytime, a guy dressed in regular clothes, walking briskly up to the front of the house and ringing a Ring. "But over 95% of break-ins actually occur in broad daylight, which is why I invented the Ring Video Doorbell." The homeowner, a woman sitting at an outdoor café over pastry and coffee, answers her phone. "Hello?" The sketchy fellow at the door says, "Uh, I'm doing free tree-trimming estimates?" She answers, "I'm bathing the children right now," and watches the would-be burglar back away and scamper off, his face already recorded.

I was an awkward TV pitchman at first. But Karni was good at loosening me and everyone else up. It wasn't surprising: As a scrum half on the Babson rugby team, he had to control the pack, move people around.

When I got stuck, he asked me what my persona was. "Don't think about it, just say."

"Inventor," I said.

"Okay, and who's the greatest inventor on TV?"

"James Dyson."

James Dyson was not just the inventor of the best vacuum cleaner in history, someone who famously went through 5,127 "failures" before he reached his goal. He was also a very cool English gentleman with a great accent who sounded authoritative and elegant in his commercials.

Karni got hold of a transcript of one of Dyson's spots and had me read it. We took the essence of those commercials and used them as a guide for ours.

Over the course of my TV commercial career, whenever I stiffened or looked off, Karni would mouth "Dyson" and I would remember: *I'm an inventor, I know this invention better than anyone.* Things went smoothly from there.

In three days, we shot five commercials. Karni cut them up in clever ways, so we had dozens of 15-second and 30-second variations. Simon was tickled at what we had accomplished: He had gone from overseeing multimillion-dollar spots for Apple with more than 50 people on set to ones for under $10,000 for Ring, and he said ours were more satisfying because we had to do everything ourselves.

I hired another twentysomething, Micah Stone, who worked with a friend of mine in San Francisco, doing marketing and analytics. On a Saturday Micah flew down to LA for an interview. Sunday, we drove around Santa Monica, then had dinner in Venice, sat outside, and watched the people go by. Easiest sales pitch ever. "Dude," I said, "is this not better than Silicon Valley?"

"Yeah, Silicon Valley sucks," he said after a moment. Monday, he gave his employer two weeks' notice. Micah became my media buyer and analyzer.

We looked at the commercial timeslots that were available, usually because they were not primetime, and bought cheap ones, like those after 11 p.m. We also saved by buying at the last minute, when the prices collapsed even for some better slots because no one had claimed them. It's called "scatter media": Your slots are all over the place because you're buying whatever's left. We tried gaming the system by checking weather reports in various cities: If it looked like Kansas City was going to have wet weather on a Saturday, we'd buy extra time there the night before, because we knew people would be more inclined to shop that weekend. We tested different lengths of commercial, different narrative hooks. For every commercial, we alternated "end cards," the image you see in the last few seconds. One card read "ring.com"; the other read "ring.com and these fine retailers—Home Depot, Best Buy, and others." The former ran on weekdays, encouraging viewers to shop online, while the latter ran on Fridays and weekends, encouraging shopping in-store (while still including our URL).

There was a boomerang effect: The retailers gave us more shelf space because of the commercials; the commercials brought them more traffic and interest in our product; they gave us even more space and "preferred placement." When a retailer sees a product being advertised on TV, they want to make sure it's front and center in their store, as if it's their own. They become attached to it. We were benefiting not just because of the metrics but the human element, too. The retailers loved that we were promoting them.

We saw a noticeable uptick in online sales during the week. On weekends, we didn't care about getting—or even want to get—all the business to Ring.com, because the volume in the retail stores was so good, and the more business we pushed to them, the likelier they were to give us even *more* preferred placement. We spent money to sell doorbells while also getting other value—in this case, the good graces of the biggest chains. *Brand, brand, brand.*

I used the logic of the small, financially strapped company that we were, and also of the much bigger company we were not. When you're a small company and every dollar matters, you need that advertisement to translate directly to sales because you need the money. Direct-sales businesses do this. Household names, on the other hand, advertise more for brand recognition, not direct sales. If you're Toyota, you advertise to be top of mind so that every 7 to 10 years, when the viewer is preparing to buy a car, they buy yours.

We had to do both at once. Many marketers think you can't do both at once; they break their marketing budgets into "sales dollars" and "brand dollars."

I beg to differ. It's *all* sales dollars. Any spend that doesn't sell your product is just a waste of money. Dyson, for one, had done it brilliantly for years—selling vacuum cleaners *and* establishing themselves as the best in their industry.

I wanted to do that for us. I just needed all the money we were now spending on TV to truly earn itself back, and then some.

O

I was invited to a gathering of tech friends with experience in direct-to-consumer marketing, who were trying to figure out how they could sell better and track better. The CMO of Zip Recruiter, Allan Jones, was there; so was my friend Scott Marlette; as were reps from other local companies. Micah came, too. There was a growing wave of marketing channels for tech companies to do TV, radio, and mixed media for the purpose of acquiring customers directly. Today we think nothing of TV spots for internet companies but back then everyone but the very biggest players were feeling things out, deciding how much of their marketing budget to put in TV—one of the reasons I was getting such pushback about my spots.

The talk turned to formulas for measuring the getting and keeping of customers—CAC (customer acquisition cost) or LTV (lifetime value).[8] One of the attendees talked "payback time."

---

8   At this moment, it's all ROA—return on advertising spend.

"Our payback time is four to six months," he said. "We acquire a customer for, say, forty bucks and six months later, tops, we've recouped the forty."

"Well, I can acquire a customer for a hundred thirty-six"—I said, quoting my cost to develop, manufacture, and ship—"and the next day get an order for two hundred."

"Is that right?" Scott wondered. "You're only getting sixty-four back from that two hundred and it cost you, what, a hundred to market it on TV? You're acquiring customers for way more than you're making on them. Doorbells are not sticky enough to justify your cost. I don't know the average number of doorbells that people buy, but I'm guessing something like 1.01? You're gonna go out of business."

Scott is incredibly smart. Did I mention he was employee #20 at Facebook? Whenever I couldn't work out a problem, I often called him and we would walk around the neighborhood or go to dinner and talk it out.

And he was right: Our trajectory was unsustainable.But Scott's logic did not apply to me and Ring. Not because we were so special, but because the only thing that would lead us to victory (solvency, then profitability) was becoming the dominant brand, not just for people who needed a doorbell (we'd be out of business in another year if that was our market) but for people thinking about ways to make their home safer. This is all we needed:

- When people thought about home security and neighborhood safety, they thought *Ring*.

- When they heard "Ring," they right away knew what we stood for.

If we were a brand, *the* brand, that would mean way more volume, which would mean a lower unit cost. And to win on brand, I had to pour as much money into that goal as possible, *now*. Later, we'd be dead. Take the big risks early.

"The only way to get the price of the product down is volume," I told the group. We were about to launch the subscription service for cloud recording, so our customers' videos would be stored in the cloud for months. I had changed my mind on the price. I'd already decided that $3 a month was more like it, or $30 per year, because everyone has the right to feel safe, not just those who could afford it. My dream attach rate—maybe 20%?—was very optimistic. But that's about what we thought Dropcam was doing. And our service was better.

The key was to get as many people on board as possible, because it would bring in more money in the moment and also meant a longer customer journey with us. It would increase the possibility of selling them other, non-doorbell products for their safety. And then the more people who had our doorbell, the safer neighborhoods would become, and a tipping point might be reached. "Brand, then volume," I said to the group. I wasn't certain I was right—my batting average, as I noted earlier, was probably around .100, though part of that was because I took swings all the time. Still, I needed to show confidence in my vision. Someday, I wanted "Ring" to be a verb that referred to *us*. Like Google and Uber. Like Photoshop and Velcro. Sure, Ring already *was* a verb…

Scott wasn't buying my bluster. Outside after the meeting, he said, "Jamie, I'm concerned for you. I don't care about my investment. I'm worried about you running out of money." It was genuine. He knew that if Ring went out of business, I would be out of money, and Stephanie, Scott's wife, was not going to let the Siminoff family live in the poolhouse.

"I'll make it work," I insisted. "Full steam ahead."

All this TV advertising was not only increasing our sales (though we were still losing money per unit) but generating excitement among my incredibly dedicated Ring team. It wasn't that long ago that we'd all been spending a serious chunk of our waking hours responding to customers who were posting obscenity-laced reviews, trying to keep them from bailing on us.

We were selling so much product online that we added videos to explain them.[9] When visitors to the Ring.com page played a video, there was an insane conversion rate. CAC, LTV—who could top *that*?

But we still needed an unbeatable presence in stores. And if the retailer put our product on the wrong shelf, or didn't honor a promised endcap, or gave it a sad-looking display, or our competitors' products showed better than ours, then we weren't going to sell. The in-store presentation was the retail equivalent of a Google search; a bad display was like having your website listed halfway down the page of search results or, worse, not even on the first page.

I hired Ollie's basketball coach and the coach's friend, two young guys looking for jobs, to build a "Ring Explorers" app: Consumers all over the country who bought and loved Ring products would "own" our display in their local store. In exchange for discount codes on future Ring products, they would make sure our display in, say, Best Buy store #XX in Milwaukee was tidy and clean, they would take photos of the displays and post them on social media, or maybe post flyers in the neighborhood. To me it seemed like a brilliant, if slightly crazy, idea: *Hey, neighbor! Buy a Ring and turn into a de facto sales rep!*

The network soon expanded to hundreds of Explorers. The guys built a whole backend for the app.

It didn't work to help sales. Take lots of swings and you're going to miss a lot. Maybe it should have worked—all the Explorers wanted to help, each in their own way. Some of them yelled at store owners to give us better placement. Others moved product around on their own. They were all well-meaning.

I still think it's a great idea, in line with what we were already enjoying: Ring users sharing their videos on YouTube.

---

9   This was the early days of product videos. They would come to be known in the industry as... explainers.

But Ring Explorers didn't work. Just move on and try the next thing. Keep swinging.

O

We recorded radio spots. We tested male voices versus female. The more data I had, the more I could convince the skeptics among my investors that I knew what I was doing by blowing all their money on marketing.

In the end, we used my voice, because it was the one we could afford.

And we finally began to charge for the cloud-recording feature. *Wired* reviewed it:

> When I first installed and set up the Ring, I had convinced myself the video recording feature wasn't something I'd need after the free period expired. But I can say now, the peace of mind in knowing any activity at my door is being recorded is worth the minimal fee. Not to mention, you can download any of the videos to your mobile device for easy sharing with family members, or in the hopefully unlikely case where it's necessary, the authorities.

One Saturday, Yassi, the team, and I went around the Wilshire Park neighborhood in central Los Angeles, offering to install Ring doorbells free to anyone who wanted them in a community about a half-mile in radius. Everyone wanted them. By 2 o'clock, we were out of doorbells; we had 50 takers (around 10% of the whole area) in a neighborhood that got hit, on average, with more than a burglary every two weeks.

Since a little after Erin's first reaction more than three years earlier, I had premised my reputation and the company's on the idea that it wasn't just a feeling of safety. I believed that perception and reality aligned because our doorbell actually *did* make you safer.

Finally, I would see if our mission was based just on hope, or truth.

CHAPTER 10

RUNNING RINGS

We were starting to win, in big ways and small.

More and more, Ring neighbors were posting videos, including one that caught a bicycle thief in the act (the camera had turned on because of the motion sensor).

There were genuinely nice moments: One morning, Yassi came to show me a photo on her phone of a Ring display in a Costco that her dad had taken, then one of him and her mom standing proudly in front of it, smiling.

But my friend Scott was also right: Solvency is good. Really good. We were a success, but we were not successful. Our cost per product (not including marketing spend) was down to approximately $100, but I needed to get us down to $80 by summer if we wanted to stay alive.

We were also losing. I was burning through money as fast as we got it—faster. All for totally legit reasons—product development, marketing—and I remained conscious of every dollar we spent. But still.

The beast was hungry. The beast needed to be fed.

O

A couple of months earlier, at a cocktail party for startup founders and investors on the rooftop of Gjelina, a restaurant in Venice, I'd met Kevin Dunlap, a former mechanical engineer who had once worked at NASA's Jet Propulsion Lab before moving into venture capital. (In other words: He'd transitioned from working with really, really smart people to just plain smart people.) When I found out that Kevin's fund was affiliated with a privately held homebuilder based in Los Angeles, I practically jumped him. Didn't it make sense for him to work with us, maker of doorbells? Think of all the housefronts! I used words like "synergistic" though I didn't really care what they did; I just needed money. Kevin tried to sneak away by saying he needed to refill his drink, but I chased him down. We were not in the doorbell business, I told him, but the home-security business; our product made you, your property, your neighborhood safer.

We promised to stay in touch. Okay, *I* promised to stay in touch. Kevin still thought I was a doorbell salesman.

But weeks later, he called to say he would be in Santa Monica and he could stop by the office for a half hour.

Two hours after he showed up, he gave me a term sheet for our B round.

He understood what we were doing. As a former engineer, he could see the breakthrough we had made. "If you tried this five years ago, it never would have worked," he said, which was true. "Wifi connectivity wouldn't have allowed it. Too expensive. Mesh routers weren't popular. Most people had a single router in their home."

I nodded. A lot of the negative feedback we'd gotten for DoorBot and even the new Rings was from customers with sketchy wifi. "Your doorbell sucks," they would tweet us. Sometimes that was true, sometimes not so much.

"The whole cost curve of the components, the battery life, connectivity—all those are coming together for you," Kevin continued.

Best of all, he was excited about the story around building trust in a community.

He got it. He got us. And now here he was, ready for his firm, Shea Ventures, to give us several million dollars.

I wondered how Kevin would feel about my using a few of the millions he was committing to us to maybe... advertise during NFL games? I didn't say any of that to my new friend, though.

The firm or individual who puts the most money into a funding round is the lead. They deserve to be the lead. Kevin wanted Shea to lead, which was fine with me. I doubted I could find anyone willing to put in more than they were.

The only time you might abandon the "who leads" principle is if you land a fish big enough that the name value alone is worth it.

I was about to land one without even casting my line.

○

I had installed a doorbell for Wes Chan, a friend and former Googler who was now a venture capitalist in San Francisco. (Over the years I've installed hundreds.) Sometime after that, Wes visited Necker Island in the British Virgin Islands, Richard Branson's legendary home. Over dinner, Sir Richard, one of the world's most curious people, asked his dinner companions to share something new and cool.

"Earlier today I was able to tell the UPS guy outside my home in San Francisco where to put my package," said Wes, turning his smartphone to Richard to show him a replay of the video.

Richard was delighted. "Okay, now I know what to get my friends for the holidays."

Soon after, I received an email from Wes introducing me to Sir Richard (really, re-introducing, since we'd met on an elevator in Brussels almost 15 years earlier), which was followed soon after by an email from Richard himself. He wanted to buy a bunch of Rings for his friends.

I wrote back that I would take care of it—and that by getting these Rings for his friends, he was also helping to make them safer.

He responded quickly. "What do you mean?"

I described the ring of security, the neighborhood ring, how our doorbell already had an impressive record of interrupting and preventing home break-ins. Our exchange was all by email, while Ring people kept poking their head in my office to ask me a packaging question or a PCB question or a "should I pay this bill?" question, and I violently waved them away. *Can't you see I'm in an intense email conversation? With the elusive white whale known as Sir Richard Branson? I am Captain Ahab and I have a harpoon and I am trying to land us Moby Dick. Go!*

Richard emailed back that he loved what we were doing.

"Well, if you really love it," I wrote, "we're about to close a B round and I'd be honored and delighted to have you involved. But it's closing soon."

His last email in the exchange: "My man will visit you on Saturday."

Sir Richard—the legend I'd encountered years before, only to be denied a chance at a more meaningful connection when the hotel refused to accept my letter to him—was seriously considering investing in us. Saturday morning, I went out and bought a fancy charcuterie platter for Richard's man, Latif Peracha, the fellow who ran his venture capital arm. When Latif showed up at 1523, he looked pissed to be there on a Saturday morning, in a leaky rundown building, vetting a company that made doorbells. He had less than no interest in the charcuterie. "Fifteen minutes," he said.

I started pitching him, talking twice my normal speed. Within 10 minutes, Latif's mood completely changed. Like Kevin at Shea, he morphed from skeptic to "Where do I sign?" He wanted to know everything about us, see everything we had planned. An hour later he was on the phone with Richard advising him to invest, even asking Richard if he could put in his own money.

It so happened that a slot in the B round had recently opened up: An investor had dropped out because they'd had second thoughts about us and the competition, especially Nest, which they thought was going to "eat Ring up." *Thank you very much, fellas, for the vote of confidence!* Now I could slot in Sir Richard, someone way more buzzworthy, whose support would go a long way toward helping us. Instead of being pissed at the investors who'd bailed on us, I sent them a gift basket of Virgin-embossed swag with a note that read, "No hard feelings. Hope you guys are doing well," scheduling it to arrive right about the time that Sir Richard's investment became public knowledge.

Having a Richard Branson invest in your business means more than just money. He was admired and trusted. His interests, whether in commerce, technology, travel, or philanthropy, were leading-edge. If he did something, the business community took notice.

I called Kevin, whose firm was set to lead the round. "Listen, is it okay if we say Richard Branson is leading the round? Instead of you guys?"

Kevin had every right to be pissed. "Yeah, of course," he said, without pause.

"You know, almost anyone would tell me to go fuck myself, because you're putting in a lot of money, you were first, and then suddenly, out of the blue—"

"Jamie: It's okay. The value of saying Sir Richard Branson is leading a round rather than some rando who lives in Orange County—we get it. *I* get it. I want the business to succeed."

Sir Richard even wrote a post for his popular blog about who we were and what we were trying to do.

Can a doorbell really work as a way to help prevent crime? Introducing Ring, the smart doorbell that can help improve your home security.

The Ring Video Doorbell calls a user's smartphone when activated, enabling homeowners to see and speak with visitors, regardless of where they are. As burglars often ring or knock to ensure a house is empty before breaking in, Ring can deter them by giving the impression the home is occupied. Even if they don't physically press Ring's doorbell button, built-in motion sensors detect visitors' movement and trigger instant mobile alerts and HD video recording.

I am an investor in Ring, and while the product is exciting and its mission to reduce crime in communities is important, what I found so interesting was the story of the young entrepreneur—James Siminoff—behind it. Like many others at the start of their journey, James tried and failed to raise money for his idea—for him it was on the popular US show *Shark Tank*. Unbowed, he went on to develop the product and the business and just raised a substantial fund raising which I joined.

This passion and determination for his business is crucial for any business builder. Indeed, when investing, I think just as much about the leadership teams as the products. James and I first met by chance in a lift in Brussels many years ago (he didn't do the elevator pitch!)…

We closed the $28 million B round with Sir Richard as the lead. Post-money, Ring was now valued at $60 million. Up to then, we'd raised $9.5 million, so this was a huge leap forward. Aside from Richard there was Shea, American Family Insurance, more from True Ventures and Upfront, plus multiple angels, including Nas, the rapper.

Then I did something else you're advised not to do: put out news of importance in the month of August, when everyone is "away." We publicized Sir Richard's investment mid-month, and because there was a dip in business-news coverage, the story got loads of attention. *Richard Branson is investing in this company called Ring! This doorbell prevents crime!* For two weeks, we received wide coverage, virtually all of it glowing. *Fortune* led with the fact that my missed follow-up with Sir Richard 15 years earlier had gone from lost opportunity to golden second chance. *Inc.* also went with the "life is full of second chances" angle, highlighting my failure to close a deal on *Shark Tank* and how that turned into a great thing.

What a great summer! We even got a nice bump in sales.

The Branson funding round made people take us even more seriously. Like, *Maybe these guys are actual players?* If Sir Richard was investing in a doorbell company, then it was probably more than just a doorbell company. The investment would help us expand our vision for the doorbell and conceive of new products and features to enhance home security *outside* the home. Owning the front door was crucial, but once we did, there were other key parts to own, too.

I took a quick break to fly with Erin to Ireland to attend the wedding of my childhood friend Zach, whom I admired for always figuring out how to live large and not give it a second thought. The wedding took place in Ashford Castle in Cong, right next to the ruins of a 13th-century monastery. The fireworks display probably cost a quarter-million bucks. Dancing after dinner took place in the castle's dungeon. Zach once again inspired me to see that my big wasn't big enough.

Back at it in LA, I hired more people, including Mimi Swain, who had come from my favorite company in the world (besides our own): Dyson. She was a talented, natural marketer whose experience there could teach us a lot. One thing she had learned at Dyson, she told me, was how much people can actually love the products they buy. They usually don't know (or care) how it works or where it gets its power. But they feel an emotional attachment to it if it does what they want, does it well, does it consistently. I had learned about that kind of product-love for our doorbells from our brief, ultimately failed attempt at a national spiderweb of Explorers, those Ring-loving soldiers who were eager to visit Best Buys and Home Depots to make sure our product was being treated with dignity.

New opportunities percolated. Mimi had worked extensively with QVC, and I wanted to do more with them, hawking Rings. Maybe I would be the pitchman, and when I needed a boost, I'd just look over at Mimi, she would mouth the word "Dyson," and I'd get extra-revved.

○

And then there was football.

Micah had determined that the optimal time to run our TV commercials was Saturday morning to early afternoon: People shop Saturday afternoon and Sunday. It's no wonder Home Depot has been a major sponsor of ESPN's *College GameDay* for more than two decades.

We needed to look into buying college-football spots. But those spots cost a lot of money. *A lot.*

Micah, an avid college-football fan (Go, Badgers!), was intent on finding ways to keep our costs down. It was still the scatter-media approach. He bought spots on *GameDay* but not far in advance, when they cost full freight. As we'd seen on lower-profile stations, there were usually a few spots that hadn't sold or someone had pulled out of at the last minute. (Most of the slots that are locked months in advance for full price are bought by the biggest advertisers: national brands for fast food, beer, soft drinks, insurance, trucks, sports apparel.) "We'll buy any leftover spots," he told the ad salespeople, meaning we were cool not being able to choose exactly when our spots would run. (So none of that "we need to air during the Kansas State–Iowa State game" or "we need to air sometime between two and six p.m." for us.) The upside was obvious: Our cost was as much as 90% cheaper that way. And because ESPN didn't know, months in advance, all of the games they'd be covering (though they knew some, for classic rivalries), we might luck into a dirt-cheap price (for ESPN) for a Tennessee–LSU or an Ohio State–Wisconsin game.

Micah didn't have to go through me. He had the authority to write a big check, which he often did late Thursday afternoons. (TV ad salespeople hate working Fridays, if they can avoid it, so they look to sell off their remaining inventory before then.) We locked up spots for some unknown time over the weekend that would have cost 5x more had we booked them two months before. We were paying 20 grand, not 100 grand.

It was so worth it. We could measure the sales spikes. Our CAC for these football-loving doorbell buyers was the lowest so far for our TV commercials. While we saw a bump at Ring.com as well as at Amazon, the real explosion was in-store at the big retailers.

We also saved money on TV advertising in other ways. We were considered a "low-cost brand" because of who we were and what our commercials looked like (low production values, "tacky" end cards promoting our URL). That style worked in our favor. For one, it saved us money. (Who cared what the media companies thought of us?) For another, it meant that when our commercials played anywhere, including in a TV over a loud bar, anyone looking could see our URL, Ring.com, on-screen. Karni and I made sure that whatever drama was playing out in the commercial, a viewer would understand immediately who we were and what benefit we were selling. Even if they couldn't hear a word, they would still know what was going on and where to buy.

We had hacked better TV, for less money, without the help of ad agencies. At that point, the only other brands I knew of that had figured this out the way we had were Guthy-Renker, a direct-response marketing company that sold beauty and skin-care products, and the George Foreman Grill.[10]

We doubled our media spend. The next month we doubled it again. We kept doubling it until we were at a half-million a month. If I stood very still and quiet, I could almost hear my friend Scott, wherever he was, wincing.

---

10 Within a couple of years, other brands would employ our tricks; I heard about more than one marketing meeting where a company pulled out a book to show the agency the strategy they wanted to emulate. It was filled with screenshots of Ring commercials.

There were new TV commercials to deal with, too—but not our own. A competitor heavily funded by Silicon Valley elites, August Lock, launched a doorbell. It was news to me, because several months earlier I had shared our whole playbook with them, thinking they were only doing smart locks (you can open your front door via a smartphone app), and we talked about partnering with them. Now, they were going right after us and airing commercials to support their doorbell. *Constantly.* The style was stolen right from the Ring playbook: low-budget drama, friendly, same lighting, spokesman just like me.

It was an *Oh shit!* moment. We knew the competition was going to get serious. We didn't know if we could respond. Maybe this was the time when someone came along and finally *did* eat Ring up.

Micah monitored the metrics of their commercials, to see how badly their TV advertising was hurting us. And the damage could snowball: If consumers started to buy doorbells from August Lock, then it was safe to assume that they'd be more likely to stick with them for other products, too.

Micah brought me the startling news. There was no denying it: August Lock's commercials *were* affecting us. Each time they ran, there was a spike in people googling "video doorbell."

Guess who had the top results?

"Yeah, dude, every time they run an August commercial," Micah reported, "*we* do better."

That's right: *We* got the bump in sales. We didn't realize it then, but we had established ourselves as the verb in the category.

O

We shot an Amazon-targeted commercial: In it, I sit by a fireplace, looking like I'm about to introduce the next episode of *Downton Abbey*, and point out that the Ring Video Doorbell, with more than 10,000

Amazon 5-star reviews (we left out the "fucking" part), makes a perfect holiday gift. Then I read aloud from actual reviews. What's better than customer testimonials?

Amazon loved it. My friend there, Nick Komorous, loved it.

We made a commercial to promote Home Depot, so they would love us more and want to give us even more prime shelf space. Our team briefly debated whether we should ask their permission before shooting but then... *why*? Why ask if we can do something, at no cost to them, that could only help them?

When they saw it, they got huffy, briefly. Then they confessed that they loved it. We put it into heavy rotation. They gave us better shelf space.

It helped enormously that the promotion of Ring wasn't coming just from us, the team that made the product. We loved the doorbell videos that our neighbors were sharing on YouTube and other platforms. So many Ring moments went viral. Teens kissing in the doorway. Dogs zooming around the front lawn. Assholes in cars speeding by (!!). Ring customers reveled in the clips, and those without Ring doorbells loved them, too. As Mimi had said about her experience at Dyson, people can love their possessions, and our neighbors seemed to love their Ring doorbells. Simon was right: The device itself provided storytelling magic. And our customers were only too happy to share it.

Our doorbells were perfectly positioned to provide great video clips (that also happened to have "Ring.com" on them). They were being shared on all the social media platforms—Instagram, Facebook, Twitter—that were exploding. For us, it was a lottery ticket. More like a thousand lottery tickets. We had content; their distribution promoted our brand, exponentially.

Maybe we hadn't reached the "breakaway speed" that every startup dreams about, but we were getting there. We had 18-wheelers double-parked on the quiet street in front of our office, accumulating tickets as they waited to be loaded with product. We were blowing away even our rosiest sales projections. I was absolutely going to lose my bet to Diego and owe him my car.

We were building a brand, articulating what made us who we were. What we were at our core. After Richard Branson's blog post, we were now ready to go full throttle on the mission, speaking more publicly about it. It's what was authentically us. That was key, because most successful brands stand out at first for their authenticity. They might not be quite as authentic after years and years of success, but at the start, they get a toehold because they're authentic. Take Nike. Phil Knight was authentically passionate about sports. They had the waffle sole, the design, the swoosh, "Just do it," yes—but what they had, more than anything, was authenticity. New customers could feel it.

(Another decision of mine that didn't work out as well as I'd hoped: Our tagline was "Always home." I loved it. I thought it was perfect. I still think so. I hoped it would be our "Just do it," but it never caught on.)

If you're not authentic and in it just for the killing? You can sell bullshit for a while. Not forever.

We leaned into what was authentically Ring: the mission. Make neighborhoods safer.

Meanwhile, all of the current and soon-to-be-former leaders in home security were just plodding along. Same script, same products. I was worried every morning that I would check my phone and see that one of them, finally, had figured out what we were doing and decided to pummel us. Incredibly, they never caught on. Or didn't care to.

The *New York Times* reviewed popular products for the smart home in a story in which we, alone, were spared ("The Pitfalls of the D.I.Y. Connected Home"), though they also highlighted the continuing technical challenges of winning in an immature marketplace.

Toward the end of an incredible year, Don, our head of sales, and I flew to London to launch Ring in the UK. Over martinis in the bar of my favorite London hotel, Dukes (yes, Don and I shared a room with two queen-size beds; Erin joked that Don got to sleep with me more than she did), I asked him what he wanted as a bonus. Don's motto had been "Go sell shit" and man, he and his team had sold a lot of shit, in the US and also the UK, Europe, Australia, and Latin America. He and his merry band of sellers had taken us, in 12 months, from barely more than zero in sales to $45 million. (Bye-bye, Land Rover.) We were in over 8,000 stores, including more than a thousand each of Home Depot, Best Buy, Walmart, and Target. We were in hundreds of Lowe's, Sears, and Brookstones. We were in every single Meijer. We were in Ace Hardware, several Bed Baths, even a few Nebraska Furniture Marts.

Don sipped his martini, considering my offer. "I would like... a Rolex Daytona and a bonus check for twenty-five grand," he said.

"Why don't you just want fifty grand?"

"You could give me that," he said, "but my wife won't let me buy a Rolex."

The week before Christmas, I walked by his desk, handed him the check for $25,000 and set down before him a stainless-steel Rolex Daytona, a sweet replacement for his gold watch that was three sizes too big for his wrist.

He had earned it.

CHAPTER 11

THE MISSION

It seemed as if everyone in home security was doing it wrong.

Let me start that again.

It wasn't that most of the people in home security were doing it wrong; they just weren't really focused on innovating in home security. They were in a different business, not home security. What they did they did well, and many of them made lots of money—but it was more akin to a *real estate* business. Their focus had transformed from security for businesses or homeowners to trying to sign and monetize long-term leases. When they hit a threshold of leases, or contracts—say, 10,000—they bundled and sold them, like real estate (so they're happy), to a financier who bought them at discounted cash flow (so *they're* happy). A whole financial market existed around this, and still exists. These "home security" companies have a job to do, and it isn't just to do better security or to build new products to stop crime. It's to get more leases.

But wait: Wouldn't these companies get more leases by doing better security? Probably, but the home-security space had been built with out-of-date technology for the last 50-plus years. In their day, the leaders in the field had been disruptors. That's how they established dominance in the first place. That's how they attracted lots of customers and made lots of money. But they got complacent. I was reminded of something Jeff Bezos noted several years ago about Sears, the once-massive department-store chain. They were the Amazon of their day, he observed, but they eventually went bankrupt because they stopped obsessing over customers. They stopped looking outward.

That seemed equally true of the players in home security. They had been best in class since a hundred years ago and slowly lost their way. They still provided some benefit on security, just nowhere near what it could have been had they stayed focused on that and not those leases. They didn't get what was now possible to make people feel safer.

We did.

So, no: The big players in home security weren't doing home-security wrong. They were just focused on doing something else right.

The fact that so many of these companies weren't doing home security the way it should be done allowed space for someone who *was*. We aimed to be the most effective home-security company in the world. And if that led us to being the biggest one in the world? So be it.

Our approach was straightforward, the daily struggle to answer a simple question: If you were trying to prevent crime in the home, what would you build? How? Where? If you were trying to stop crime in a community, what would you build? Forget everything else. Forget even whether people paid you for the product or service. Just answer this question: What would you build so that people would look at it and say, "That makes me feel safer"?

Turns out, the first product you would make looks a lot like the Ring Video Doorbell. We had turned the alarm system inside out.

Your front door is the most important part of your home. By adding video to the doorbell, we had simply made what is unknown and potentially threatening outside the home into something known, whether you were inside or not. And since everyone understands what a basic doorbell does, there's not a lot of explaining necessary to get to a better doorbell. There's almost universal pre-awareness of this thing that you're simply making better. And it took almost no time to mount and sync our doorbell.

What else would you build? Probably a floodlight camera, because you want more eyes and ears around the home. Fortunately, most homes already have lights and power exactly where you'd want to put the floodlight camera: in the power junction box on the exterior walls. So there's pre-awareness for that, too, since homeowners and business owners already understand that that outlet exists for safety and security. Our device, with its wifi-connected camera with motion detection and audio, all connected to the smartphone, just maximized the safety factor.

That "pre-awareness" factor was one of our biggest advantages. Our potential customers didn't need to imagine a new world. It was this world, just safer. If you came across an ad for a movie called *Fighter Pilots*, you might go see it but you wouldn't know exactly what you were in for, so you might not. If you came across an ad for *Top Gun* with Tom Cruise, you'd be much more aware of what you were getting and more likely to see it. (Unless you don't like Tom Cruise, but most people do.) Pre-awareness makes it so much easier to connect to people and to sell to them.

We were never a doorbell company. We were a "making neighborhoods safer" company whose first product was the doorbell. If we were trying to build the best doorbell company, we would have built the best doorbell and then stopped. Mark Cuban couldn't make that leap. The price of our doorbell sounded like a lot for a doorbell, but really quite reasonable for home security. Few people would budget $200 for a doorbell, but how about $200 for home security? If that $200 lowers the chance of someone breaking into your home? As for the recurring cost of a subscription to store your video for six months in the cloud, we were glad to settle on $3 a month.

All of that allowed a "doorbell company" to be a better home-security company than all the home-security companies out there. They didn't compete with us, at least at first, because they failed even to realize that we were in the same business.

O

If we didn't run out of money, I was sure we could win. We weren't going to start selling leases to make money. In fact—and I mean this—we didn't focus on the output, the money. (I don't know how my VCs will feel about that admission.) We focused on the inputs: the products, the customer service, the mission. If we did the inputs well, the output would take care of itself. As I noted earlier, I told the team that we didn't "sell." We got *rewarded* by our neighbors when they understood what our products did for them, which was make them feel safer. They rewarded us by buying our products or using our services. It wasn't some semantic trick: We genuinely believed we were being rewarded.

When we finally turned on the subscription service in 2015, it changed everything: We had a new recurring source of revenue—and from the start, it was attaching at a higher rate than we expected.

I felt uncomfortable: I wasn't used to such good news that was just... good news.

The service also brought us some of our greatest reviews. One neighbor wrote, "Thanks to video recording, I caught my wife cheating on me. Wouldn't have happened without you guys—BEST DEVICE EVER!! FIVE STARS!!!!!*****"

The content exploded the viral marketing and community interaction that Ring neighbors were already sharing with the world, now that they had more video saved. The majority of videos were of strangers, perhaps would-be criminals, backing away or sprinting away when they realized they were being recorded. But there were silly and sweet moments, too. One fan from Canada tweeted her Ring video content daily, always to the tune of "Ring My Bell," the disco song performed by Anita Ward.

Come for the security, stay for the fun.

O

Ring's ability to help fight crime was loved by so many neighbors and police departments throughout the country, but some people misunderstood how our doorbell and video recording worked. They thought our cameras represented an invasion of privacy. What we really did—as good technology should do—was make things more efficient. Previously, after a burglary, a cop went door-to-door in the vicinity asking if anyone had seen anything or had video of the time period in question. Extremely labor-intensive for the police, and nerve-wracking for private citizens who now had a cop on the other side of their front door. With Ring cameras up and down the street, a neighborhood ring was created, and the police or fire department could send out an alert/request to those in the area of the break-in, requesting video from those whose doorbell cameras might have detected something important. The owner of such video could do whatever they wished in response. That homeowner decided who, if anyone, got access to their video. If they didn't share it with the police, they remained anonymous and no one knew they had been asked. But most people wanted to help and shared useful video with the police, joining in to be part of the solution.

We evolved our messaging so that we were actively referring to three rings of security: around your front door, around your home, and around your neighborhood. We were doing our part to impact crime, and now each of our neighbors was, too.

There was lots of misinformation floating around on social media platforms that Ring gave police free access to all the video saved in the cloud. We never did that and never would.

O

Product quality is always necessary but sometimes not sufficient. There were lots of companies out there, including ADT, Amazon, Google, and a growing list of startups, that either already knew or were beginning to comprehend just how massive and untapped the home-security sector was. (The *Times* article that had roasted our competition noted that there were now "hundreds of security webcams on sites like Amazon that promise peace of mind for under $200.") And with the increasing confluence of favorable factors—very available capital at low interest; smartphones; improved video streaming with connectivity; the explosion of the App Store; the smart home becoming a known, accepted concept; another leap in chip complexity; a lot of the technical advances that Kevin Dunlap had pointed out—we at Ring needed to be the first, if not only, wifi video doorbell that people thought of.

We were not going to slow spending. We were dropping a million a month on TV commercials. I hired more engineers. When would we reach a threshold where I could take a breath and say, "The job is done"? Would I actually recognize it when it came?

Results were in from the doorbell test in Wilshire Park, the neighborhood in Los Angeles where we'd installed Ring doorbells free on approximately 10% of the homes in a concentrated area to see what effect, if any, they had on crime. After six months, the verdict from the Los Angeles Police Department:

Burglaries in the neighborhood had decreased by 55%.

In a sense, this was the first KPI—key performance indicator—our company had had since its founding that we were truly succeeding at our mission. If the test had shown no impact, then all we'd been doing for the past couple of years was putting plastic and printed circuit boards on people's homes for no reason.

The data confirmed it. Now we knew we were doing it right.

CHAPTER 12

KLEENEX®, BAND-AID®... and RING® (and SHAQ®)

The only thing better than collaborating with Shaquille O'Neal— NBA Hall of Famer, constant TV presence, and arguably the most recognizable and universally liked person on the planet—was collaborating with Shaq without having to explain to him who we were and why our product was worthwhile.

It all happened because of a Super Bowl commercial we didn't do. There was almost no way we were going to do one, to be honest, given the incredible expense and how militant Micah and Simon and I had been about keeping our TV commercial costs down by hacking the system— buying unclaimed spots at the last minute, buying at odd hours, buying spots in places that were predicted to have Saturday thundershowers, using me and our backyard for shoots to save on production costs, and so on.

Still, what company doesn't dream about doing a Super Bowl commercial? And if we *were* going to do one, then we needed a celebrity. I called a big Hollywood agency to float the idea, and during the course of the conversation I punted on the Super Bowl fantasy and decided instead just to see about celebrity endorsers. Fame alone didn't do it for me. Given how mission-driven we were at Ring, I needed a celebrity who authentically cared about what we cared about—neighborhood security—and maybe, somehow, was even known for it.

The agent sounded hesitant. Probably he was skeptical that a company our size needed an A-lister, or could afford one. But he promised to gather some leads.

A week later, he called back excitedly. There was only one person who fit the bill perfectly: Shaq. He already knew about Ring and loved our product. Recently he'd bought a big house in Atlanta and called a local security company to estimate the cost of protecting his new home. They priced it at $80,000, which Shaq thought was a gross overcharge. On a subsequent trip to Best Buy to purchase TVs and other electronics, he came across the Ring Video Doorbell, bought it, installed it himself, and said it "worked perfectly" (huge sigh of relief). A while later, while in China, he got a notification on his phone that somebody was ringing the doorbell at his Atlanta home. As he started giving directions to the delivery man from the other side of the world, he thought, *Damn!*

So *he* was interested in *us*! It got better. Not only was Shaq a customer and a fan who owned Ring doorbells and had given them to his circle, but he had expressed admiration for law enforcement and security his whole public life. His stepfather was a career army sergeant. Shaq himself had gone through the Los Angeles County Sheriff's Department Reserves Academy, become a reserve officer with the Los Angeles Port Police, trained to become a Miami Beach reserve officer, and been sworn in as a sheriff's deputy for the Clayton County (Georgia) sheriff's department. (Undercover work was probably off the table for him.) He'd spent real time and effort helping to make neighborhoods more secure. If there was one celebrity who was pre-authentic to our brand, Shaq was it. We could not have asked for a better candidate for a possible partner.

The agency set up a meeting for us in January at the upcoming CES in Las Vegas. I didn't tell any of my team that Shaq was scheduled to stop by our booth. Had Yassi, the world's most rabid Lakers fan, heard he was coming and then he didn't, she might have quit. Then again, I was concerned that when he did come by, she might have a heart attack.

I was out in front of our booth (more turf, higher waterfall, more trees) when Shaq, trailed by gawkers, introduced himself, the world's least necessary introduction. He asked for the CEO.

"Hi, I'm Jamie," I said, my voice sounding funny.

"How you doing, Jamie? Nice to meet you. I'm waiting for the CEO."

"I'm the CEO."

He looked me up and down, disbelieving. He said he was expecting a distinguished 60-year-old with gray hair and glasses. He told me we had a great product, that he had 10 Ring cameras around his property. He said this was his first home without an elaborate professional security system installed by others. Thanks to Ring, he had protected his whole house for under $3,000.

The more we talked, the more I realized a partnership could actually happen. He said he would do commercials, and we would pay him in equity. (Done!) Then he smiled that incredible, sly Shaq smile. "Seriously, if you need me to help you get the word out," he said, "I'm available."

When Yassi came by, she nearly passed out.

If some of my investors had been less than thrilled about my spending close to a million on the domain name and millions more to air TV commercials on college football Saturdays, they probably weren't going to love my spending a lot on Shaq, even if it was equity. (They did not.) At my next board meeting—I finally had an actual board—one of the attendees objected strenuously, the same person who had most strenuously objected to our TV commercials. Fortunately, Adam from True Ventures and Kevin from Shea backed me up. "This is the biggest no-brainer of all time," one of them said. And since Shaq wanted to be compensated in equity, it was very doable.

It was a smart move for Shaq, too. The gentleman I'd bought the Ring.com name from could have learned a thing or two about finance from the NBA Hall of Famer.

O

I had sold product on the Home Shopping Network and now had a chance to do it on their archrival network, QVC, too. At the time, QVC was in 100 million American homes. Like HSN, it was a tremendous platform from which to address and educate the consumer directly, nationally, *live*. I brought Ollie with Mimi and me to West Chester, Pennsylvania, just outside Philadelphia, where QVC Studio Park is headquartered. We got there an hour before my appearance to meet the host and get prepped. Though I was always comfortable appearing in public in just a blue Ring T-shirt, jeans, and running shoes, Mimi, who had a lot more experience with these platforms from her time at Dyson, had bought me a Ring-blue suit from Nordstrom in the nearby King of Prussia Mall.

I joined the host on-air for 10 minutes for a morning "Special Value" segment. I showed viewers how the doorbell worked, talked about what made it different and better than anything currently out there, and answered questions from people who called in. I loved it. The only more direct contact you could have with potential customers was standing in front of your display in a retail outlet or going door-to-door. When my morning responsibility was done and the host took over to promote and sell Rings for the next several hours, Ollie and I found a nearby swimming pool for the afternoon, then returned to the studio after dinner for another segment with me and the host pitching and bantering.

At first I had been against the steep retail discount that QVC offered. Tom Czar and Dan Forde, who owned a firm that represented brands to the network, explained that "QVC" stood for Quality, Value, and Convenience, with the V the most important to their customers. "If you're selling a product that's the same price here as Home Depot, Amazon, Lowe's, what's the selling proposition to the QVC customer? Why would she buy?" Tom pointed out. In those first two pitches, we even threw in an extra warranty and access to a QVC-dedicated 800 number for their customers: *Have a problem with your Ring doorbell, call this number, we know you purchased from QVC, and we'll take special care of you.*

I loved doing the spots, and not just because over time I got to be good friends with Tom and Dan, or because one of their clients was Dyson, the company I admired so much, and the reason Mimi knew them. I loved doing it because the pitch was true mission-driven storytelling, not just the transaction of selling. I'd pitched our proposition so many times, for so many different audiences, but really the message was the same whether I was selling one doorbell to one customer or trying to get millions of dollars in funding from a venture capitalist. On the QVC spots, I explained the what and the how, never talking down to the audience about the technology but keeping it pretty basic (most people don't want to know about their wifi network; they just want everything to work). I especially loved talking the why—and QVC, always live, was perfect for that. My pitch was simple and universal. Everyone loves their family and their home. That's it. For almost all of us, those are the most important things. Whether you're in your 20s, your 40s, your 70s—doesn't matter. Those things matter to everybody.

Given the large discount, though, selling on QVC might not have been the most profitable venture. If I'd had a traditional CFO, or a traditional C-level executive of any kind, they probably would have tried to talk me out of spending time doing things like that. But if they had, I probably would have fired them. Because it would mean that they didn't understand what we were doing. We weren't just after what's called a "first-order effect," but a second-order one. The first-order effect of the TV pitching was making money from the selling of doorbells. Yep, we were doing some of that. But the second-order effect was conquering territory. And we were doing *a lot* of that. Once you own territory, it's hard for your competitors to get it away from you, unless you do something really, really stupid.

O

With all the sales we were racking up, you would think we weren't still bleeding money. You'd also think we now had a nice office, or at least not a dump.

You would be wrong, on both counts. Our building was still grimy, with air-conditioning that never worked when you needed it, and much of it looked like one big storage space. Bean bags were flung about. We had taken over the mezcal office, but there were still too many people for the square footage. And everyone was still working their ass off.

There was one meaningful advantage to having such a crappy building. A man from Dallas contacted us at one point, claiming he and his partner had patented a process that had been used in our doorbells. We didn't believe it was true—or, if it was, that it was a sue-able offense. We invited them to come see us. They flew to LA, walked into our office, and looked around at the splendor, and you could tell it was not what they were expecting, or hoping for. Maybe a mouse or rat skittered by, or they felt a drop from the leak in the roof because it was time for me to tape some new garbage bags up there. The men showed us what they had patented; we still felt that we had done nothing wrong or actionable. But I knew we didn't want to get tied up in a legal case. Neither did they, I could see. They threw out a very reasonable number and we settled then and there. I'm certain that after spending a half hour in our dilapidated office, they felt lucky to get anything.

Being underestimated can be a very good thing.

We'd been doing serious business long enough, with enough partners, that it was probably time to hire another lawyer. We had one who had worked on patents, but I needed more help. Leila Rouhi joined us from a mergers-and-acquisitions firm; before that, she had worked in fashion, at a company that sold $400 jeans. In our interviews, she said she wanted to be part of a team. At her job in M&A, her role was mostly transactional: Do a deal, get to know people on the other side really fast, then never hear from them again—"or when you do, it's because something in the deal went wrong." I told her that our mission was to make neighborhoods safer. I told her about our rings of safety, how we were democratizing security, bringing it to people who couldn't afford it, because every family had something they wanted to protect. Ring wasn't a doorbell company, I stressed, but a company trying to make the world safer. I told her I thought she was at the right place. She was hooked. She was ready for a new challenge and a new purpose.

Leila had been with us barely a week when she informed me that the manufacturing agreements we had with several partners in Asia had never been executed, just drafted. Which was more than you could say for some other partnerships, where we had no formal agreement of *any* kind, just a handshake. She helped tighten those. She was a lightning-quick study who figured out what needed to be done without needing to be told. She hadn't been with us a month when we received a search warrant from local law enforcement: Someone in nearby Long Beach had been murdered inside a home with a Ring doorbell, and the police wanted to check out the video to see who had entered and exited the home in the hours leading up to and immediately after the crime. It was the first time we'd gotten such a request, though not the last. Leila worked with Jason Gluckman, the engineer who had built the system to process all Ring videos, to track down the footage. Adhering to the search warrant, we shared it with the police.

They found the killer.

O

Big things were happening. Josh Roth and his team launched the "Neighbors" feature I had thought up for the Ring app, allowing users to post video streams from their doorbell cameras of dubious characters peeking around their homes, or someone stealing a package from their porch. I worried about how our neighbors would feel about the feature;

what if your neighborhood only had a couple of old posts on it? I wanted to get bigger before launching it. Josh was certain it could be a huge hit.

He was right. Almost the moment the feature went live, neighbors used it. And loved it. A great lesson, like the video doorbell: When you're an improvement over something that doesn't exist—not a very high bar—people may love it. I had an idea of what the feature *could* eventually be, and that was the problem. I would have gotten in the way of letting our customers decide what it *would* be.

We raised a Series C round of just over $60 million, led by the legendary Silicon Valley venture capital firm Kleiner Perkins; our valuation was now just over $200 million. I could tell some of the Ringers, doing the math, realized how much money they were worth at that point. Not too shabby...

... yet I still found a way to lose money, month over month. How was that possible?

We absolutely murdered the top line (sales)—we were moving product briskly, both the main doorbell and the other products, too. I expected our sales to more than double in the coming year versus 2015. The video storage subscription service was exceeding our not-so-modest expectations.

But we also absolutely murdered the bottom line (expenses). I was frugal with any chance I saw to save, yet I wasn't afraid to spend big if I thought it was needed. The beast had to be fed. And though we were great at guerilla marketing, sometimes you just have to spend big—for the perfect domain name, say, or the perfect celebrity endorser, or anything you think can give you the edge in credibility, attention, and—always the last of the three with me—sales.

○

It was time to declare war.

Business can seem so hard when you face a first-time problem, which seems impossible to solve, but then you solve it and things are good—until you face the next problem, which seems even more impossible, but you solve that one, too. Sometimes, though, the problems can get overwhelming and it feels as if you just can't solve them fast enough.

We were facing challenges on several fronts. I had to do something dramatic.

From: **Jamie Siminoff** <j@ring.com>
Date: Thu, Mar 17, 2016 at 6:57 PM
Subject: Going to War
To: All <all@ring.com>

Team,

Next Friday the 25th of March is going to be the day that Ring officially declares war. Over the next few days all of you will be getting camouflage t-shirts (they look awesome) and I would like everyone to wear them to work on the 25th.

What and who are we going to war with:

We are going to war with anyone that wants to harm a neighborhood. We are here to "reduce crime in neighborhoods" and we are going to make that happen. So, to the dirtbag criminals that steal our packages and rob our houses, your time is numbered because Ring is now officially declaring war on you!

We are going to war with copycat competitors. We are seeing more and more of these hacks announcing that they are going to come into our market and take what we have built. Well, I can tell you, they have no idea what they are in for because this team is here to win and we will crush anyone stupid enough to come into our path. So, to the people trying to take what we have worked so hard to build, we will take you seriously and not underestimate you but, in the end, we will crush you because you cannot win against a team that is as dedicated to winning as ours!

Now that you know who we are at war with, what do we do to win:

1. In war, secrecy is key to winning. We have spoken about this before but the old term of 'loose lips, sink ships' is very true, so please be extremely careful with any info you learn when you are talking to people. It is imperative to our survival. Also please remind all vendors of this. We see vendors sharing info with competitors all of the time.

2. In war, it is won by a team. Our team needs to work together and be dedicated against a common cause and each member has to always be on their game or it can break us all. I believe we already have this and the competition coming in only makes us stronger.

3. In war, you have to make sure that you use your resources carefully. We raised a lot of money but we need to make sure that we spend it wisely and on the right things. It is not about winning each battle but it is about winning the overall war. We will win both the battles and the war but we need to remember that this is going to be a long fight.

With all the great things happening at Ring, it is very easy for us to forget that there is a war going on out there. We need to stay focused and make sure we continue to execute. I am not sure if we will ever be in a comfortable position but I can tell you that the next 12-18 months are going to be critical in our becoming a real success.

I am proud of the team we have built. We are an amazing group of people and I know that with the right focus we are going to be able to WIN THIS WAR!

Jamie

PS: If you want to wear war paint, army helmets, etc., feel free to on March 25th. Just no guns. That would be taking this a little too far:)

To help our people blow off steam (or maybe it was primarily for me), we gave out foam bats that looked like they were wrapped in barbed wire—replicas from *The Walking Dead*, the TV zombie hit. Feel free to slam them against walls and desks, they were advised. (Please, not each other.) We dressed in camo. I gave out dog tags; the number on your tag was your employee number. (We broke open our typical badges, removed the electronic guts, taped them to the back of the dog tag, and that became your work badge.) We drove military-style six-wheel vehicles down the streets of Santa Monica.

We were not fucking around.

Once Shaq agreed to be Ring spokesman, I called Karni to help us figure out exactly what promotional assets we wanted to create with him, how many, how much. Shaq would come to LA and give us his time and effort for a certain number of commercials, cutdowns, photos, and social videos. Karni had worked with top business clients and celebrities, so I trusted him to know how to budget this kind of thing. He emailed me the plan and tally. I called him back seconds later.

"Jesus Christ. A private jet?"

"Shaq's a big guy. I'm sure he'll ask for one."

I couldn't argue with the logic. We calculated that it was cheaper for our whole crew to go to Atlanta, where Shaq lived, than for him to come to LA. Just before Mother's Day, we all flew out there. I drove around the neighborhood with Shaq in his special-edition Dodge Charger, with the back seat removed and the driver's seat extended all the way (so he was more or less driving from the back seat). We stopped at several homes to install doorbells for whoever wanted one. We drove all over the city. We had lots of product for Shaq to give to neighbors—single moms and others curious about the Ring doorbell. We visited higher-crime neighborhoods, installing more Rings. Karni had cameras rolling the whole time, collecting footage for commercials and social media spots.

I wasn't fully prepared for what unfolded. I don't think anyone could be. Wherever Shaq went, everything turned magical. He was like the sun. Yes, he was a celebrity and yes, he was a gigantic human being. But he was also beloved by pretty much every segment of the population. No one did not know Shaq. Everyone wanted at least to stand near him. Have him hug them or kiss their baby. The world turned into one big knot of paparazzi. He was literally the Biggest Star in the World, but also a total what-you-see-is-what-you-get regular guy.

We finished earlier than scheduled, and when we got back into his Charger he said, "What do you want to do? Need me for anything else?"

"We could go to a Best Buy and you get a Ring for your mom for Mother's Day," I said.

Karni and the camera crew followed. There was no way we were allowed to just go into a store and film a commercial without corporate clearance. I figured we'd try without permission and ask forgiveness afterward.

On the drive over, we talked about crime nationally and crime in Atlanta. Shaq mentioned a young man who had recently been killed by gunfire, and that he had paid for the funeral anonymously because if he said anything, then it became about him. The more we talked, the more I realized: When a good deed is done anonymously, there's a not-small chance Shaq is behind it.

We pulled into the Best Buy lot. It turns out, when you're with Shaq, there are lots of things you can do that you couldn't otherwise.

As we approached the entrance, everyone looked. There would be no "quietly walking in." Trying to be unobtrusive with Shaq is like trying to hide an aircraft carrier in a small marina. Pulling a baseball cap low over his eyes doesn't work. The camera crew trailing us made it obvious that we were filming something. (Had the iPhone camera been as good then as it is now, then one of us could have filmed the whole thing surreptitiously.) The security guard bolted toward us.

"Shaq! Can I get a picture?"

Shaq didn't just buy a doorbell for his mom; he started buying all the Rings in stock—for customers, for *their* moms. He paid for a television that one man was about to purchase. Everyone was taking photos, videos, texting, calling friends so Shaq would say hi and prove they weren't making it up. By the time we left, the store was packed.

In his car afterward, I asked, "Do you ever get tired of that?"

"My mom told me you can be the guy that wants to meet someone or the guy people want to meet," he said. "I'd rather be the guy people want to meet. So I'm always happy to do it."

Karni put together the commercial, we sent it to executives at Best Buy, they spent a morning losing their minds over our filming in their store without permission, then told us how much they loved the commercial. We aired it constantly.

○

As we grew, we heard here and there about civil-rights groups that worried about how our doorbell might infringe on people's privacy. Yet we never heard that from our neighbors. As we went around with Shaq, the main queries people wanted answers to were "Where can I get one?" and "Is it hard to install?" and "I have two/three/five entrances to my home, can I get two/three/five Rings?"

We also filmed in a studio. Shaq had been doing professional endorsements since the age of 17, he'd been in movies, been a TV presence for years, played basketball in crowded arenas with millions watching, so he knew exactly how to be when the camera was rolling. He could do everything in one take. I could not. He got me to relax in a way that even Karni couldn't. Mostly it was his sense of humor. He kept calling

Karni "Spielberg." We were pinching ourselves. We posted a photo of Shaq and me on Facebook. We got the first Shaq commercials on the air within a month.

He was #1 and that's what we Ringers needed. No offense to Dr. Pepper, Avis, and Burger King, but #2 or #3 wasn't good enough.

For our sector, we had to be Coke, Kleenex, Band-Aid. Dyson. Apple.

O

In early April 2016, the Los Angeles Police Department held a press conference to talk about ways that new technology could help them fight crime. They devoted part of their time discussing crime reduction in neighborhoods with more Ring doorbells. I thought of the possibilities for insurers to get wise and lower premiums for homeowners who used Ring doorbells and our other devices.

It was a triumphant moment for our company, and no one deserved praise more than Yassi, who had directed the Wilshire Park campaign from the start. As a thank you, I got Yassi and her boyfriend (now husband) two tickets to a Lakers game. It was the last game of Kobe Bryant's career. I paid a lot for the tickets, and the seats weren't even that great because everyone wanted to be at that game. Yassi Yarger, the world's greatest Lakers fan, had to be there.

The fans expected a special night. I doubt if anyone knew that Kobe was going to score 60 points, the greatest regular-season cap to a career, ever. Maybe Kobe knew.

"The best night of my life," Yassi told me the next day, and for weeks after.

O

Yassi was less thrilled with me when she got flagged at an Apple Store in Manhattan, after another tech show where I made her buy Macs to return the next day, rather than pay to rent the conference's inferior audio-video equipment. Along with CES in Las Vegas, we now presented at other conferences around the country, like Pepcom and ShowStoppers. After the dozen or so times that Yassi had done my Mac trick, Apple's system finally identified her as a serial returner. Before they would take back the computers this time, the store manager came out.

"We notice that you continue buying Macs and returning them the next day," he said. "Can we ask what you're using them for?"

"I work for a company with lots of new people starting all the time," said Yassi, totally spitballing and hating me for putting her in this position, "so we get them Macs... and many of them don't, you know, work out."

They 100% did not believe her, but they took back the Macs, adding the built-in 5% restocking fee. She had gotten to be world-class at packing them up. Back at the hotel she told me, "I can't do this anymore."

"Why not? We're not renting gear from the venues. They suck."

"I hate lugging those things in a taxi across town. We've raised tens of millions of dollars in funding and you're telling me we still can't afford an eight-hundred-dollar rental?"

She had a point. But I never made my people do anything that I hadn't done myself, many times. Every dollar saved was $100 saved.

O

Diego teaches courses at USC's business school, and he invited me to come speak to his class. I was delighted when one of his students asked, "Why didn't you work harder to have a patent so Ring could block all its competitors?"

I asked the class, "By a show of hands, if Ring had such a patent, do you think we would be bigger than we are now, or smaller?"

About 90% of the class voted "bigger."

"I'm almost certain we would be smaller," I said, "much smaller. Because if we had such a patent, we wouldn't have the constant fear of competition in us. Without that, we wouldn't have the constant need to innovate. Intellectual property is great, but it can also make you complacent. Fear is the ultimate motivator. You can always count on fear. I'm driven constantly by fear."

Another student raised her hand. "Travis Kalanick came to the class to speak to us last week," she said, referring to the co-founder of Uber, "and he said he has no fear."

I smiled. I *wish*. "Well, good for Travis," I said. "He's obviously very successful. It just goes to show that everyone's different. Personally, I'm a barbell of confidence and terror. I don't know where I'd be without fear."

○

Nextdoor was a well-funded, fast-growing Silicon Valley social media startup for neighborhoods. With Ring videos turning into ideal viral content, our customers and especially their engaging content seemed perfectly suited to the platform, which made Nextdoor a great place for us to advertise. I agreed to pay them millions over three years to advertise there if we were granted an exclusive: No other home-security outfit got to advertise on it.

Given how much money we were guaranteeing, and that no one else had even advertised yet on Nextdoor, they agreed.

Simon, now my chief marketing officer, threatened to quit when I told him I'd just taken millions out of the marketing budget to advertise on a network that no one advertised on. Pointing out that we had an exclusive didn't make him any happier.

It turned out to be a great billboard for us.

I truly never knew whether our big swings were going to hit or miss. (How can anyone, really, until they do it?) I was terrified that this one might not work out, and thrilled when it did. I was even more thrilled that Simon didn't quit.

In conclusion: barbell of confidence and terror.

O

August was turning 30 and deserved a nice birthday lunch in Beverly Hills. I had Erin join us because I was giddy about the gift I had for August. He had meant so much to me and the company from the beginning.

At the end of the meal, I pulled out an Apple bag I'd hidden under my chair and handed it to August.

"I got you an iPad," I said, with a big, dopey smile.

"Jamie, stop," said Erin.

"What?" I turned to August, excitedly. "Open the iPad! Open, open!"

He opened the bag and saw a green Rolex box. I could see him tearing up, and I was too, and Erin was a mess, but August had yet to realize what was actually inside. It was not a brand-new Rolex. It was the one my father had given me for my college graduation.

I'd had it engraved:

**To my
other son
on his 30th
Jamie**

It was one of those things that feels right the instant it comes to you.

When August could finally manage a few words, he asked, "Don't you want to give this to Ollie?"

I hugged the birthday boy. "By the time he'll want to wear a watch," I said, "I think I'll have a few others I can give him."

O

We always had our board meetings at Nobu. When I'd first established the board, it was just three of us, so going to an expensive sushi restaurant wasn't that crazy. But as I raised more money and added members, the meetings became more expensive. I wouldn't change to a less expensive place, though, because I'm superstitious. (See: Don and I sharing hotel rooms wherever we traveled.) Things were going well for the company. We were in the midst of Ring War I, fighting enemies on various fronts, but we were killing on sales. A quarterly board meeting meant I had people who had actually invested and believed in my vision and in me— so why mess around? And the sushi was incredible.

To some of the people around that table, it looked as if I took risks, maybe even stupid, crazy ones. But I never saw it that way. I was honestly not a gambler. I approached everything in what I thought was a rational way. If risk was called for, then that was the logical, non-risky route to take. If risk could be lowered, then I would do everything I could to lower it. Gambling, to me, was just stupid.

So in mid-2016, when we contracted a team of hundreds of engineers in Ukraine, mostly in the capital city of Kiev, it did not remotely strike me as a gamble. They could help develop the main Ring product over there, at reduced cost. We already did much of our cloud streaming in Argentina. To lead the team, I hired Jason Mitura, who in 2008 had started a successful company in Ukraine—Viewdle, which enabled embedding of computer vision technology into smartphones, and was acquired by Google for millions. He was incredibly smart and competent and knew his way around. My own experience of eastern Europe from years before, and of overseas generally, told me there was what Jason called a huge "perception arbitrage"—that is, the perception from the US was much different from the reality on the ground, so you could build a big advantage by taking the "risk" of building over there. Most of the workers had a certain grit that I loved and recognized, an us-against-the-world sensibility. *Wait, you have a weird accent and haircut and went to a school nobody's heard of? Who gives a shit? You have the potential to be the best electrical engineer I've ever met.*

So much talent over there. Even better, so much opportunity. I loved it.

O

The Ring Floodlight Camera came out on October 16, 2016.

Ring had begun expanding into international markets, but we still had so much territory to protect at home. Even with all the expansion in our ranks, especially in customer service, many of whom were now based at a huge call center in Phoenix, my team was fraying. I was, too. I had once told the *Wall Street Journal* "it takes seven years to really build" a startup, and I believed it, but with that comes a big problem: It's really, really hard to work like a maniac for seven years straight. Maybe in the Bible; not in 21st-century America. I prided myself on my tenacity, a willingness to keep working despite pain and suffering. On never, ever stopping. But even I was having moments where I was losing a little faith and looking for an easier way out. You can push human beings so hard for only so long. I understood that we were going to make mistakes; mistakes were okay. I could feel us, all of us, possibly edging into the dark world of excuse-making when things got tough.

In late October, sadly, I felt it was time to declare war once again.

Team,

It is clear that as we have grown bigger that we are in a very dangerous position again and back in a full and much bigger war this time.

On Friday the 28th we will officially be going back to war, Ring War II (RW2).

Here is what we are fighting for in RW2:

Our Neighbors- We are no longer delivering to our Neighbors (what we call customers) the level of product, support, and uptime that we should and that they deserve. This is unacceptable and if we do not improve on this, we will lose. Just this week we had an issue on Android that cripples the Neighbors use of our service, resulted in poor ratings, and it was handled without any care to the impact it was causing. This will no longer be accepted.

Ourselves- We are now in a position to win, but the other side of that is losing will come only because of our lack of execution. I am seeing this throughout the organization, we have team members killing themselves for us to win and others are just not taking it that seriously. Ring is not a job, we are here to win, to fight and to make EVERY neighborhood safer.

So what do we do from here:

PRODUCT- We need to make sure we focus on making our products better and more reliable. Our products must live up to the customers' expectations and we need to work harder than ever before to produce better software and hardware.

NEIGHBOR SUPPORT- Our neighbors need to be blown away every time they interact with us. We need to give them fast solutions (ONE CALL AND YOUR PROBLEM IS FIXED) and have them leave every interaction wanting to tell their neighbors about our company.

INNOVATION- We know what to do and we have the capabilities. Now we need to execute and deliver. We can show the world what community security is at scale but if we do not work hard and execute, someone else will beat us to this.

Global- This is no longer a US company or a US fight. We are global in team and have to be global in sales. We have to win everywhere or else we will allow a little no one to become a big someone.

Rings of Security- In RW1 we had only one Ring of Security to focus on, the **Ring of Security around your front door**. We now have products in a second Ring, the **Ring of Security around your home**. And before the end of the year, we will be launching the **Ring of Security around your neighborhood**.

Stop Using Excuses- Ring is suffering from a disease of excuses. I keep hearing "We are too busy to do X." The reality is we are doing less with more people today. For those who were here when we transitioned from DoorBot to Ring, you have seen what we can do. We relaunched an entire company and product in 9 months with a few dozen people total, while running the business. Today we are over 500 people. So if you cannot do something because you think you are too busy, take a hard look at what you are doing and get more done. If someone is creating a bottleneck for you, blow it up, be creative, work around it, come to me, but do not let it stop our progress! Be efficient, work harder. Winning this war against real competition requires sacrifice and focus, excuses will not matter if we lose.

We won our first war and it took the same concentration and focus that RW2 will take. We can do this. It is fully in our power to accomplish this but it will take everyone giving everything they have got to the cause.

I thank each one of you for being part of this team and know that united and dedicated, we can accomplish any goal.

NOW LET'S GO WIN A WAR AND CHANGE HOW
PEOPLE LIVE IN NEIGHBORHOODS AROUND THE
WORLD!!!!

Jamie

PS- As with RW1, feel free to come to the war party
in camo, fatigues, army helmets, etc., but no guns/
weapons. That is taking it a little too far:)

PPS-As with the first time I sent this email, war is
an analogy to the fight that we have as a company
against the forces that are against us. It is meant to
stand for the significance of what we are doing and
also create something as a team that we can all join
around and fight for. It is not meant in any way to be
taken too seriously. We should have a little fun with
it but not at all take away from what the servicemen
and women do to protect all of us in our respective
countries.

Harsh? We had a job to do. We'd gotten ourselves into such a good
position—pole position, if you're a car-racing fan. What a shame it would
be to crash out.

O

There was a sea of good news.

I was out of town for work on Halloween, which bummed me for
multiple reasons: I hated to miss trick-or-treating with Ollie. And it was the
holiday that, around our office, we called "the Super Bowl of doorbells."

Erin, Ollie, and friends trick-or-treated in our neighborhood. She
called to tell me that house after house after house was outfitted with
Ring doorbells.

"I think this might work," she said excitedly.

I loved it. I also wasn't going to celebrate.

Then, another fortunate break. We were still spending a lot on commercial TV airtime, still hacking the system by waiting for the last minute with national sports programming so that we paid a fraction of the price for placement. We had lucked out before, landing cheap spots late in close games between ranked college football teams, but never like this:

We had bid on an open slot for a World Series game—open because it would take place during extra innings, which are less likely than not to happen. It was 80% off the full price *if* the slot materialized.

Then in Game 7, the Cleveland Indians, who had not won a World Series in almost seven decades, scored three runs in the bottom of the eighth inning to tie the Chicago Cubs, who had not won a World Series in more than a century. The ninth inning was scoreless, so the 6–6 game went to extra innings. We were guaranteed a spot after regulation and before the end of the game. That meant lots of eyeballs, for not much money, and it yielded one of our biggest sales days ever. (In the end, the Cubs barely held on to beat the Indians, 8-7, in 10 innings.)

Maybe the biggest and therefore best good news of all was the signup rate we were getting for the cloud subscription, our most reliable source of recurring revenue. At first we thought there had to be a mistake. I prayed that there wasn't; that it was real. It was a nice change from lying in bed at night, wide-eyed, praying that what *was* real—like, the amount of money we were losing—was not. This time I was *rooting* for the reality.

Our subscription signup, or "attach," rate—the percentage of our customers willing to pay $3 a month or $30 a year to have their video saved for six months in the cloud—was incredible. Like, absurd. When Micah showed it to me, I thought it was an obvious error.

The rule in direct-mail marketing: A response rate of 2% is considered excellent.

In email marketing, a good response rate is around 10%.

Survey-response rates hover between 5% and 30%.

Nest's Dropcam camera rate, it was reported, was somewhere in the 20% to 30% range.

Ours? Higher. Way higher.

When we looked further at the number, we saw that it crossed all demographics. Paying the monthly fee had nothing to do with your income or wealth, your gender or age. It seemed that $3 a month was the sweet spot: Anything higher was insulting; anything lower wouldn't enable us to get to solvency within the decade. It felt like we had won the World Series.

O

---------- Forwarded message ---------

From: **Jamie Siminoff** <j@ring.com>
Date: Fri, Nov 25, 2016 at 5:33 PM
Subject: Black Friday Battlefield Report
To: All <all@ring.com>

Team,

If Ring is at War, then Black Friday is one hell of a big battle! What it lets us do is talk to customers in store (the average store sells 1-2 Rings per week) but today in 3 hours at Best Buy I was able to sell almost 10 units and talk to about 20 potential customers.

A few interesting learnings:

1. If someone asks after the Black Friday Sale is over to match the price, the answer is "The Black Friday $99 'door buster' sale was only on a limited number of units and sold out fast. We're sorry but that was a once-in-a-lifetime opportunity for that price and was here and gone in a flash.

2. People are still very scared of Rings being stolen (every person I spoke to said this – 100%! – which is crazy). We need to talk more, in everything we do, that we guarantee that if a Ring is stolen,

we'll replace it. This was the #1 reason that people seemed not to want to buy.

3. People want more cameras. Huge opportunity. We need to make sure our app continues to evolve to make multiple cameras an amazing experience.

4. We are a strong brand in the store. People saw it and focused in on Ring. This business is definitely now ours to lose and that's something that can still happen. We need to make sure we ship new products on time and that we make every experience with our neighbors awesome.

5. I did have a current neighbor come up to the aisle. I asked them if they knew about the Ring. He kind of frowned and said that he already had one. When I asked him more about it, I realized he had wifi distance issues and needed a Chime Pro. We need to be proactive with the customers out there that are not having great experiences. People buy our products based on what they hear. We need to make sure that everyone with a Ring is having an exceptional experience and the Chime Pro is key to that.

It looks like the holidays are going to be busy, and supporting the new neighbors after Christmas will be a huge strain on the organization. We have a month to prepare so I ask everyone to get focused and do what we can to keep winning!

Jamie

So many customers who got the $99 Black Friday deal emailed to thank us for lowering the price so they could afford it and they would "now feel protected, just like everybody else." I flew to Phoenix to rally the customer-service team at our call center there. For every new group of hires, I made sure I was up there in front of them, sharing the company's mission, the key points of our culture. Turnover in customer service was—is—traditionally very high, so I did all I could to have the people who were handling customer calls feel as invested as anyone. I said to the agents, "So many of the VCs who turned us down told me, 'Hey, China'll knock you off. They'll make the same doorbells at a third of the price. How can you compete with that?' And you know what I say to those VCs now? 'You know how we're winning? By selling an *experience*, not just a product. By giving them incredibly useful features that no one else has. By creating a caring community among our neighbors that no one has built like we have. By having the best customer service and the truest mission anywhere. That's how!'" I felt like I was giving a halftime locker-room speech.

We "created" a holiday, National Package Protection Day, two days after Cyber Monday. With so many people buying from Amazon and using its two-day delivery service, we figured that lots of crooks would be roaming American neighborhoods that Wednesday, stealing packages from porches and easy-to-breach vestibules. So many of our neighbors posted videos of thieves being thwarted from stealing packages. Local TV news stations ate it up, re-broadcasting the best ones. We were creating a virtual national neighborhood.

Our Ring Floodlight Camera was selling very well, as was the Ring Video Doorbell 2 with Adjustable Motion Zones. But the basic Ring Video Doorbell was our bread and butter.

I'd found a great assistant (though I hate the word "assistant"), Limi Abdusemed, whom I deeply trusted and who'd clinched the job at one particular moment in her interview. I asked about her background and she said that when she was growing up in an immigrant family, there were only two acceptable routes her parents wanted her and her sisters to take: Be a doctor or a tennis player.

"Oh, you have great tennis players in your family?" I asked.

"Zero."

"Then *you* must be very good."

"I'm terrible."

Pre-med hadn't worked out either, and she found her way to us. Hired. The tennis world's loss was my gain.

The company was growing by leaps and bounds, now surpassing 700 employees, including hundreds in tech support and over 200 great contract engineers in Ukraine, which allowed us to develop more products and gain even more momentum.

That all sounded good—but something was seeping in. When you start getting too big, too chaotic, and maybe even too pleased with yourself, that's when you have to really watch out. We were a success. We still had yet to be successful.

Two things were clear to me, as they had been from the beginning:

1. We had the opportunity to dominate in this incredibly important and lucrative space.

2. As long as we focused on what we were doing and what we *could* do, the competition coming at us didn't matter, even as we knew they were out there, everywhere.

I needed everyone at the company to understand this, which is why I kept letting them know it.

Because make no mistake: The real war hadn't even begun.

CHAPTER 13

IT'S ALWAYS 50/50

Half full, half empty, you choose.

Insane growth. We hit $170,000,000 in sales in Y2016. No typo there. Three video-camera products out, four in the pipeline: floodlight camera, outdoor camera, new video doorbell, high-end doorbell (the Pro). Many of our customers were adding a second device, often as many as 10 or more. The ridiculous subscription-attach rate continued to be ridiculous. The Neighbors feature on the Ring app was killing.

If we were still short of "breakaway speed," it wasn't by much.

Own the front door and you own the home.

The doorbell, it turned out, was just about the ideal way to attract new customers to try other home-security technologies and make our brand literally a household name. It was the entry to the home and to ancillary products. In fact, the first outdoor camera we launched was really just a Ring doorbell without a button. So much for Mark Cuban's "I just don't see the progression."

That wasn't all.

I believed we could almost triple our sales in 2017. A crazy goal to anyone outside of 1523, but we believed, or at least I did. Yet that also meant—here's the constant conundrum with hardware—that we'd be ordering close to *a quarter-billion dollars' worth* of components, from multiple vendors that required payment well before Q4, which was when 50% of our sales happened. It was scary, when I stopped to think about it, so I tried not to stop and think about it. But it was basically like buying a $5 million house when you're making 100 grand a year because you're betting that you'll be making a million a year soon enough. Really? How can you know that? You can't. Past performance does not guarantee future results.

But as a hardware company, we had no choice, because one thing is not debatable with hardware: You cannot sell what you do not have. If you didn't order it, there's nothing to sell. So we were always on the knife's edge. What if some event occurred before the holidays? Market crash, earthquake, zombies? Life is more surprise than not. What if Google or some upstart came along and crushed us? Had we established ourselves as the go-to brand deeply enough that we were impenetrable? No, not yet. We were losing $5 million a month consistently. We kept hearing that competitors were coming out with their own video wifi doorbells any day. Suddenly we're not talking zombie odds.

Wait, there's more! Before I go half-empty, there's more half-full highlights. While out in Vegas for a trade show, I met the QVC sales director of electronics for dinner at a tapas restaurant in the Aria Resort & Casino. He told me, "If you come to QVC exclusively and leave HSN, we can do monster business together."

"Great!" I said. "Let's do a hundred million in retail sales in the next twelve months."

He paused. His idea of "monster" was probably more like $10 million. But he nodded and we shook hands.

So much positive news. What on earth could possibly go wrong for us?

I was in talks with Stuart Miller, the CEO of Lennar, one of the largest home builders in the country, to put Ring doorbells on all their houses, at no cost to them or the homeowner. If the homeowner wanted to pay the monthly subscription later on, great. If not, no problem. (The stats told me that most would.) Either way, the homeowner got a handsome state-of-the-art video doorbell. And Lennar bought into our vision: that having a Ring video doorbell next to the front door gave homeowners an elevated sense of safety and security from the day they moved into their new house. It was an ideal partnership: Lennar, visionaries in their field, led by a founder who cared deeply about the neighborhoods he built, could say they were working with an up-and-coming smart-home tech company. And Ring got to embed itself even more as the industry standard. Of all the founders I worked with, Stuart may have been the one most aligned with our mission.

We were also developing an alarm in-house. If our mission was to reduce crime in neighborhoods, and if our goal to further the mission was to provide full security for the home, then an alarm made sense. But we thought we could make a better security system by adding different monitoring "modes" to our cameras. Are you home alone and feeling vulnerable? Your camera or cameras would be in one mode. Are you about to have a party with lots of people coming in and out? Your cameras would be in another mode. Are you on vacation? Another mode. By adding these modes, plus a supporting basic alarm, the growing Ring ecosystem would make our neighbors feel even safer.

Did I say there was more good news? There was. Amazon was about to release their new Echo Show product and invited us to be one of their first partners, a presence on the screen.

There's more—$109 million more. That's how much we raised the first month of the year, in a Series D round of financing led by DFJ Growth, Goldman Sachs, and Qualcomm Ventures. Amazon's Alexa Fund was in, too. True and Upfront were in. Everyone wanted in! We had to turn people down. Quite a change from my early money search for DoorBot.

We had enough to reach breakaway speed.

Meanwhile, I was pushing my team to be as dedicated, hardworking, and kick-ass as they had always been, something I had been praising them for ad nauseum for more than three years. I wondered, though, if the message was chafing. I thought of those NBA head coaches who have a few years of success, take their team to the playoffs a couple of times, even win a round one year, but then get fired because it's clear the players have tuned them out after hearing the same pep talks, the same tantrums, the same "We'll get to the mountaintop" speech year after year. To win, to dominate, we had to deliver an Apple-like experience, a Dyson-like experience, to every one of our customers, every day, with every product. Every new feature had to be as close to perfect as possible. In every single encounter our neighbors had with us, they had to come away smiling, their problem fixed, so impressed that they couldn't wait to tell their friends about Ring's awesome tech support. Fucking 5-star reviews up the wazoo. I believed we could do all that before Nest (via Google) or anyone else, for that matter, came out with their video doorbell, and thus blunt any possibility of them unseating us.

I wondered if I had driven everyone too hard, too long. Myself included.

I thought of "perception arbitrage," the idea that Jason Mitura, the head of our Kiev engineering team, and I had about eastern Europe: the difference in judgment between those who've actually been there and those who haven't. When you're there, you see chaos and poverty, sure, but also vitality, opportunity, talent, action, an energy different from anything in this country (but similar to other places I'd worked where people are desperate to make it, like Kinshasa and Tunis). The rules of the road are different and you have to be there to recognize them, adjust to them, learn them, thrive despite them.

Those who haven't been to those places or had those experiences make judgments based on what they read, videos they see, comments they hear, and that's all valid—but it's usually the "off-the-shelf story." It's nothing like actually being there. And the delta between what it's really like, for those who are in it, and what it appears to be like, for those who aren't, is massive. There's money to be made in that difference. Serious money. The perception arbitrage.

It's very different being in the arena versus watching from the stands.

There was a perception arbitrage between what was true about Ring and what wasn't. Maybe the biggest problem was that even its founder was unsure which was which.

Everything was going our way. That would continue, I believed, as long as we never took our foot off the gas.

O

At a Ring board meeting, I got into a shouting match with members who wanted me to raise the monthly price of the subscription.

Their reasoning? *Hey, if we're getting this many people paying the monthly fee of $3, why can't everyone pay an extra $1 a month?*

An argument I found to be—what are the words?—oh, yeah: fucking idiotic.

There was a threshold for people, especially ones for whom the extra dollar a month made a real difference. Not for any of us sitting around that table, eating sushi at Nobu. But yes, there were actually people out there for whom it would make a difference. What about the possibility that we might lose a bunch of them because we could squeeze everyone else?

I finally relented. "You can increase the monthly price to whatever you want," I said, "if I'm not in the company. Or if I'm dead."

But maybe I was wrong. Maybe my focus on democratizing safety and security for as many people as possible actually hurt us in the long run. I was all about the mission. But what about the business? Maybe if I'd agreed to a small increase in price, it would significantly boost our bottom line, thus securing our long-term independence. I was so dead set against appearing even remotely elitist that in doing so, maybe *I* was the fucking idiot.

O

I got a call from the head of Zonoff, an engineering company focused on "Internet of Things" devices, including alarms. We were partnering on product development and licensing their technology for a smart alarm of our own. They had about 80 engineers and other staff working in Malvern, Pennsylvania.

"We're shutting down," said their CEO.

"Wait, what?"

They were out of money. They had done a deal with ADT, and recently Apollo Global Management had taken ADT private. It seemed that continuing the operations of a business that was building DIY alarms for companies such as Ring was not helpful, so they closed the spigot for Zonoff. It was hard to blame anyone; had I been in their position, I'd probably have done the same. But I hated seeing so many good Zonoff people, many of whom I considered friends, out of work.

"When is this happening?" I asked the Zonoff CEO.

"It happened."

"I mean, when does it take effect?"

"It already did."

"Holy shit... okay, book a conference room at the Hilton Garden," I told Zonoff's boss. "Tell everyone to be there. I'm taking the red-eye. I'll see you in the morning."

The conference room at the Hilton Garden in West Chester, Pennsylvania, was overflowing with former Zonoff employees, some of whom had babies or young kids in tow, since they hadn't had time (or maybe money) to get child care on such short notice. I knew many of them. "Everyone, your salary starts today, same salary, same everything as yesterday," I announced. "Only difference, you're a Ring team member now."

As an inventor, I saw a problem and wanted to fix it. Here was an opportunity for us to keep working with some very smart people. I was also getting the source code for the work Zonoff had been doing for us;

a provision in our agreement with them stated that if they ever went bankrupt or ceased operations, Ring had the right to the technology they had created for us (we'd spent more than $1 million on it), including the source code. I went to the Zonoff office and got what I thought was legally ours.

Then I called the landlord of the office and told him I wanted to rent it. Same rent, same occupants. Cool? Cool. The now new Ring team returned to the office later that day.

Having done all that in under 24 hours, I thought it might be nice to inform my board of directors.

> BOD,
>
> …
>
> As of last night Zonoff has ceased operations as their deal with ███████ failed to come to a close.
>
> I am now in Philadelphia and we are going to hire the team and set them up…
>
> If we are able to pull this off, still need to fully on board the team, then we will be getting an incredible asset, team, and control over a core part of our business…
>
> *Jamie*

Soon after, an executive at ADT called to inform me that I could not do what I had just done.

"Can't do what? Hire people that were not currently employed? People who were looking for jobs? I'm confused, what can't I do?" I was relishing acting like an asshole. Right makes might, I thought. "Which part?"

"Jamie," he said, trying to reason. "Don't do this."

"Tell me why."

"We won't let you."

I paused for maybe a nanosecond. "Really?"

It was one of those moments where, looking back, the smart move would have been for me to hang up, or at least say I'd call him back in five, or take a deep breath, or walk around the block, or go for a nice, long run in the woods. Best of all would have been to get on the phone with Erin. She'd know how to talk me down. Or, if I was at home, I would have sat with Ollie while we watched a Lakers game together on TV.

Why didn't I just get on the next flight to Fort Lauderdale, meet with the ADT guy in person, and figure it all out? *Why?* I always did that. In-person *always* brought better results.

But no, I had to do it this way. ADT and Ring had had a relationship, a decent one. For years they had been selling our products and services. I was setting that on fire. Right makes might, doesn't it?

"You know what?" I began. Any speech beginning with "You know what?" probably won't end well. "Can you just do me one favor? There's a bunch of furniture in the Zonoff office that's yours, and it's getting in the way of everything we're doing, so could you send someone to come get it out of my office?"

Then I hung up.

I obviously didn't mean to put myself, my dream, my business, and everyone I knew and loved and really cared about in jeopardy.

But I just had.

Late on the last day of April, my trusted lawyer and confidant Leila answered a call from lawyers based in Delaware with whom we sometimes worked. They informed her that ADT was suing us for illegally using their intellectual property in the alarm we were building.

Later that night, I let the team know.

From: Jamie Siminoff <j@ring.com>
Date: Mon, May 1, 2017 at 12:29 AM
Subject: ADT declares war on Ring, picks the wrong company to bully
To: All <all@ring.com>

Team,

Tomorrow, May 1st, ADT will officially sue Ring. They are suing us for a number of reasons. Becoming a ████████████████████████████ ███████████████ our mission to reduce crime in neighborhoods, ████████████████████████████ ███████████

Please read the below for background on the facts surrounding this lawsuit and some behind the scenes info. This will become our official response if needed for now, though it is not public.

There will be much more on this, as well as Ring War 3...

Jamie

I told my team the claims were baseless (though I probably should have pointed out that I was not a lawyer). I told them we fully expected to prevail. And that ADT may have thought they could distract and intimidate us but they were wrong. We were not letting anyone push us around.

I was trying to keep up the team's spirits, but my upbeat talk was at least as much for me.

I knew just how badly this lawsuit, bullshit or not, could hurt us.

Like, end us.

O

One of my favorite books is *Walt Disney: The Triumph of the American Imagination*, by Neal Gabler. It's also one of the longest books I've ever read or listened to: I consumed it across 36 hours while flying, driving, and running with my dog. (I prefer listening to books to reading them.) What I loved most was how, by the end, you realize that Disney, a Master of the Universe if ever there was one, always felt behind and about to go under, always dreaded what was coming, was always frustrated. His story was *not* a Disney movie. It was an *anti*-Disney movie, an anti-Cinderella story. He was pretty much tortured his whole life, trying to get things just the way they were in his incredible imagination. Walt Disney always felt on the edge of ruin.

Which is what made the story so alive and real. He fumbled along his whole life until he got things right—but even when he did, creating movies and theme parks and characters that brought so much pleasure to billions, he knew there was more he could do, and it would probably lead to more torture. His sense of mission was a blessing and a curse.

That story was feeling closer to home by the day.

○

It played out like a meal for, I don't know, a third date? Fifth date? Will we, won't we? Nick Komorous of Amazon had come down to LA and we were out to lunch, sitting across from each other. We looked deep into each other's eyes. There was a little awkwardness in the air. Excitement, too.

"Jamie, I think it's time to go to the next level," said Nick.

"Wait, are we dating?"

"I think we should."

"Should what?"

"You know."

"You mean...?"

"Yes."

"I need you to say it. Do you want to put a ring on it?"

"Fine. We should buy you. Amazon should buy Ring."

Okay, the exchange didn't go *exactly* like that, but it kind of felt like it. Several years of flirting and footsie as we partnered with Amazon to sell our products, as they asked us to partner on Echo Show, as we made commercials to promote them, as their Alexa Fund invested in our latest round. It certainly gave me peace of mind to have Nick and Amazon in our corner. What consumer-product company wouldn't want the interest of civilization's single biggest online retailer as a platform for its products?

I was torn about the offer. Ring was my baby. I didn't yet know the size and terms of the offer Amazon might make. I knew that with their resources and reach, their technology-forward approach, and Jeff Bezos's long-term thinking, we could build Ring into something huge. My feeling about customer satisfaction as the be-all, end-all aligned with Amazon's. They had a development group with an impressive history of elevating startups and entrepreneurs they'd acquired, from Zappos to Audible to IMDB to Twitch. Amazon was the only large company I knew of that had bought companies like ours and consistently brought them to the next level. (There should be a museum devoted to all the exciting smaller companies bought by other giant companies, never to be heard from again. Actually, not a museum. A graveyard.)

As for our appeal to them: What could make for a better acquisition than a company that helped protect the "last yard" of countless Amazon packages the world over? The front door was the gateway to the home. Amazon had every reason to want to manage deliveries and deter theft as much as possible.

Yet even with so much alignment, I just didn't know if I wanted to go all the way with them. How much independence would I lose being part of a giant corporation? Our sales were huge, both in growth rate and absolute numbers. I'd been thinking more seriously about taking Ring public. I would probably have to raise another round, maybe as much as $400 million (more?), then engage an investment bank to manage the process. But if we kept on our current trajectory and there wasn't a zombie apocalypse, we *could* be massive, and I could also make the Ring team much richer.

Oh, and also: Ollie loved Ring like it was a family member.

There was risk in agreeing right away to Nick's overture. I had told him more than once that Amazon was the only company I would ever sell to. But if we did that, we'd definitely be leaving money on the table, lots of it.

But there was risk in *not* agreeing to be bought by Amazon. They could swoop in and buy one of our competitors, like Blink, based in Boston, smaller than us but growing fast and impressively creative. Money was obviously no issue for Amazon.

So both were gambles. One was a bigger gamble. But both were gambles.

I told Nick I was not ready to sell—but we should keep talking.

Ring finished the first half of 2017 with $149 million in sales, more than double the first half of the previous year. My hope of tripling 2016's number no longer looked like a crazy dream. We had grown to around 1.3 million households, with just under a million active subscribers.

For perspective, Vivint Smart Home, the largest pure residential-security company in the US, had just over 1 million subscribers, slightly more than we did.

ADT had 6.5 million subscribers across residential and business.

I wanted to be bigger than ADT by the end of 2018. To do that, we needed to grow by 5x in the next 18 months. Maybe I was losing my mind. But four of five US homes had no home-security system, so there was lots of room for a trusted, innovative brand to grow.

And we would have the resources to do it. I was raising by far our biggest round yet, close to $200 million. We had one investor committed for close to half of that, so they would be the lead. True and Upfront would be in. Kleiner Perkins would be in. We needed a couple more investors to close the round. It was great having suitors and partners for multiple paths (acquisition, IPO, just keeping going as is for a while). With our new products, we were set up for a huge holiday season. Amazon was keeping a close eye on our progress. It was all good. We were looking like kings of the world.

The first week of September, it happened, as I knew it would.

Nest announced its video doorbell. I wrote the Ring team that I was certain that it was probably a very nice piece of hardware, and that Nest would probably soon be announcing a security system or alarm, too, and that no one could possibly beat us but ourselves. If we stayed 100% focused on our mission of reducing crime in neighborhoods and making people's lives safer and better, we would make the best products and offer the best services to do that. We had the most affordable and effective products in that market. And—just as a reminder to my team—we should never, ever publicly mention our competitors by name.

○

Nick from Amazon called. How were things going?

I told him that I was raising another round. That's just how it goes with hardware, no matter how well you're doing. He knew it. In fact, the company he worked for, named Amazon, had been in the same boat for the first several years of its existence. Completely to be expected.

Nick said Amazon would be interested to take part in the next round. I was thrilled. They were hedging, too—looking at us as a possible acquisition, but also investing in us, which improved our chance to stay independent and perhaps be worth a lot more.

And we really needed the money soon. I asked our CFO, Mel Tang, just how bad things looked at the moment.

"We have money for three more pay periods," he said. "We have to close this round."

Oh, and we also owed about $70 million to our manufacturers. Those bills would soon be past due.

With the new college-football season underway, Micah bought commercial spots for multiple games every weekend, again getting us massive discounts by grabbing spots just hours before gametime that hadn't been locked in by pickup trucks, beer, and fast food. Our total TV spend had shot up, and now we were on more than just *GameDay* on ESPN. We found great value on CBS games, too. NBC tended to be more expensive (they had Notre Dame and some other prime games that everyone wanted), so we steered clear of it and other places where we couldn't find bargains.

In October, we got more great coverage. Our Ring Floodlight Camera won Home Depot's Innovation Award for suppliers. Then a sergeant in the Philadelphia Police Department publicly thanked Ring for helping them apprehend a convicted felon. The thief wasn't the brightest criminal ever: He went around stealing Ring video doorbells (among other things). "Your assistance allowed our detectives to secure an arrest & search warrant for our target, resulting in 7 counts of theft and related charges," wrote the sergeant. "Without your help, we would have not been able to bring closure to this crime pattern."

I flew to Ukraine to check on the team there and catch up with Jason.

The morning of November 2 was a great day for Ring. For the second time in our history, we launched a product on time: the Ring Protect alarm. Other than the Ring Video Doorbell 1, which I'd sworn to Jon Callaghan I would launch by the date I promised, every other product, going back to our DoorBot days, had always met with delays of weeks, sometimes months. Now, we were delivering a product exactly when we had told stores it would be en route to their shelves. Some units were already sitting in Best Buys and Home Depots. Others were on trucks.

Everyone on the team had worked so hard to make it happen. Since I was in Kiev, I couldn't tell the Santa Monica team in person how proud I was of the achievement.

Nick called to say that Amazon was sending a term sheet to buy us.

Holy shit! Top of the world!!!!

I told him I was thrilled by the possibility—but I was also racing to close a round of financing for around $200 million. (He knew that. His company was part of the round.)

But now the calculation for Amazon was changing. They wanted to get the term sheet to us as soon as possible because once that funding round closed, with or without their investment, we would officially be valued considerably higher than we were right then. In other words, once the funding round closed, we might escape their comfort level for buying us.

It also happened that for the first time in all the rounds of financing for DoorBot and then Ring, money would be "taken off the table" and set aside for me. I had always paid myself an unusually low salary, for reasons I no longer thought made sense. Not that I should have paid myself a lot from the start—could hurt individual motivation, might hurt company morale—but neither did I have to make some grand point by making myself one of the lower-paid members of the team. Now, with this round of financing, $10 million would go to me. Erin and I would have a level of financial security we had never known. There was a condominium where we sometimes went skiing that I was hoping to buy.

How great was all this? I had an amazing Plan A *and* an amazing Plan B. My biggest problem might be choosing which mountaintop to sit on.

I thought about my father, and how I wished he could see this. He was conservative and risk-averse by the time I showed up. He put deals together, but always small ones. He never bought himself a fancy watch. My older brother, John, followed a similarly conservative, steady path, acing his way through school, through college, earning an MBA, then working in IT at a financial-services firm.

Yet when my father was younger, he had co-owned a plant that forged steel pipes for oil refineries; it was tools from that business that I would use to take apart television sets and learn how to fix things. He and my Uncle Adam co-owned one of the country's top "trotters" (horses for harness racing), and Dad even wrote a book titled *How to Win at Harness Races.* Then he became a husband and a father and pulled back on the throttle.

Me? I wanted to do something that had never existed before. I often thought of the line from *Butch Cassidy and the Sundance Kid*: "I shoot better when I move." So I kept moving, constantly taking swings. And my adventures, frankly, made for better stories. When I was home, I shared with my father and brother my tales of global intrigue. Dad couldn't get enough. *Wait, you were installing a satellite receiver in Kinshasa and narrowly missed a presidential assassination?* Just hearing the details leave my mouth made me proud, a little thrilled, that my father knew that I was handling myself, and that he was vicariously along for the ride. For as long as I knew him, he rarely set foot outside the greater tri-state area. Maybe he had become a 5% guy, but he was smart enough to see that there were vast opportunities out there, for those bold enough.

"God, you won't believe what your brother's up to..."

John once told me that that was Dad's favorite way to start a conversation with him.

O

Between my team delivering the alarm on time and the great progress I was observing in person in Ukraine, I felt stoked about the upcoming holiday season. Most important, we were truly expanding our mission of reducing crime and making neighborhoods safer, which tipped me away from one of my two great options.

I texted Erin.

> I don't think I can sell.
> 323 people in Ukraine. So much cool stuff they are doing.

I'm down for whatever you decide

> Better for Ollie. He loves Ring.
> I just need to be more balanced going forward and chill out a bit.

Back at my room at the Hilton Kyiv, a beautiful modern hotel on a grand boulevard, my phone rang just before midnight. Jason. He said that Leila had been trying to reach me on my cell; forward-moving as Kiev was, cell service could be spotty. Jason sounded like someone who did not want to make the call he'd just made. I knew that Leila was on vacation in Iceland, so that didn't smell good either. "She's going to call you," he said ominously.

A minute later, the hotel phone rang. Leila sounded on the verge of tears.

"What's wrong?" I said.

"ADT got the injunction. We have to pull the alarm off the shelves."

The ruling meant that we could not sell the alarm we had launched *that day*. The product we had worked so hard to finish and get out on time. Future cash was now sitting on store shelves and in trucks, and a judge in Delaware had just ruled in favor of an injunction, meaning it was illegal for us to sell a single unit until the lawsuit was resolved. (And if we didn't win, then it might be illegal for us to sell it ever.) Thanksgiving was in three weeks, Christmas in eight.

Clearly, the moment we shipped, ADT had alerted the court, claiming material harm. With the confidence only a non-lawyer can have, I was sure we had the facts on our side, and the law, too, and that our adversaries were going this route to save their hide, nothing more.

Yet in lawsuits, you can have the facts on your side, and the law, and *still* lose. I'd already been wrong about the injunction. It didn't matter at all whether I thought it was a bogus lawsuit; the ruling could absolutely kill us. When anyone asks me today about lawsuits and their chances, I tell them I don't care if you're holding a bloody knife over the lifeless body. I don't care if you were in the South Pole the same minute the crime was committed at the Equator.

It's always 50/50.

"Jamie?" said Leila.

I think she thought I was going to blow. But I knew that wouldn't help. I had to be calm. Pilot mode. Both wings had fallen off, and the tail was gone, too. So? I couldn't sit and cry about it. As calm and collected as I was projecting on the phone, the phrase *You're probably going to lose the business* danced in my head. "Okay," I finally said to Leila, "let's just figure out the next steps. We'll be fine."

We devised a plan, beat by beat.

First, I called Nick at Amazon. "It's no big deal," I told him, "but just so you know, there's an injunction on the alarm. It doesn't change our numbers at all. We didn't even project it for this year. The Series E is shaping up great. The injunction is really more annoyance than anything else—"

"Dude, it *does* matter," said Nick. "This is your mess" —weirdly, he didn't sound unsympathetic— "and you have to clean it up. When you do... we'll be there."

If you clean it up, he meant, but he couldn't say that. He was now more colleague than friend. The term sheet would not be coming.

I called my contact at the fund that was leading the Series E, also just a "heads-up." *No big deal, we're still on track—*

Long pause. They said they would get back to me soon.

Hours earlier I wasn't just on top of the world; I had my choice of mountains. Now, both peaks looked like mirages.

I called Jason and told him everything that Leila and I had discussed. I told him he wasn't leaving Ukraine until his team built a whole new alarm from the ground up. We'd built stuff before; we could do it again. "You know what, that might take too much time," I told Jason. "Do that *and* also buy that alarm company in Kiev we talked about. And ask around, find the smartest engineers already working on one."

I would worry about where to get the money later. Better to be working on solutions in parallel. If the plane's crashing, why launch one parachute when you can launch three? "We need to release another alarm at CES in January," I told him. That was two months away. Not announce a new alarm; *release* it. What I was asking for was beyond nuts. But it was better than doing nothing. If I kept moving, I wouldn't think about how dire things had turned in just a few crazy, horrible hours.

O

It was well after midnight when I finally dropped onto the bed at the Hilton Kyiv. I closed my eyes, though there was no chance of sleep. I had to remain calm. To be "balanced" and "chill," as I had texted Erin just a few hours earlier. Pilot mode.

I reached for my phone and texted her again, providing her with my detailed analysis of the situation.

We are fucked.

CHAPTER 14

THE R-WORD

I had traveled something like 200 days in the first 10 months of 2017. That works out to a lot of miles. A lot of days and nights away from home. I always took the extra flight to get home, even if it meant flying out again 12 hours after landing in LA. I never thought of myself as a workaholic. I wasn't someone who put my work before my family. When Ollie wasn't in school, I took him around the world with me and we loved it, Batman and Robin. The trip to the factory in China, when the first DoorBot came off the line and Ollie had just turned 5 years old—that was one of his first memories. What could be better than being with your dad in some exciting place, watching something come to life that never was?

My biggest sacrifice was sleep. I couldn't recall the last time I'd had a good night's worth. Everything was always going so fast, I don't know how I kept up. I'm surprised I didn't make more bad calls than I did.

Then again, the one who made it go so fast was me. And it felt as if I'd just made my biggest mistake yet, a possibly fatal one.

Once we got sued, I understood what ADT thought they had on us. One of the toughest parts of the lawsuit was that it was a trade-secret case. ADT claimed that the technology used around the Ring Protect alarm belonged to them, and "the fruits" of that technology *also* belonged to them. And since the 80 employees of Zonoff who had become Ring employees possessed knowledge that ADT considered theirs, *other* products that Ring was making might be "poisoned fruit." It could even create ambiguity throughout the business. Not just alarms. *Everything.*

I didn't have a confidentiality agreement with ADT, nor did I have a restriction about hiring ex-Zonoff employees. The problem was, each of the employees had a confidentiality agreement with ADT, so we were aiding and abetting in their breaking of that agreement—or *could* have been, which in the eyes of the court might be just as bad.

I didn't have to agree with that interpretation. I still don't know if I do. I felt as if the facts were on our side. The company had closed. If I went and hired a bunch of engineers that they had put out of work, why was that my problem? As the tech journalist Stacey Higginbotham wrote on her excellent tech website, Stacey on IoT (Internet of Things), if the lawsuit "takes Ring out of the equation, it can prevent a company that has had tremendous success selling security to consumers from further disrupting the market."

Exactly.

Then again, fair is fair. Maybe there was another side to it. It's not as if I wasn't capable, once in a blue moon, of reacting without all that much deliberation. Why had I hired the ex-Zonoff employees? Because they knew how to build an alarm. And where had they learned that? While working with ADT (at least partly). Was it possible that what they'd learned was the confidential information they were not at liberty to share? I guess when I put it like that, I have to see it their way. Sure. Seems more than likely.

When I hired everyone, I did not know about their confidentiality agreements. (I realize that ignorance is not an exonerating factor.) After the injunction, I understood that those agreements existed.

Maybe the biggest thing I had to ask myself: Why did I release an alarm product just *days* before getting a term sheet from Amazon, just weeks before we were on track to close a nine-figure Series E round? Two events, either of which would have just about insulated us from calamity and allowed us finally to breathe, and see something truly great come from the years of hard work?

Was I poking the judge who had signed the injunction? Why hadn't I just postponed the alarm until after the investment round closed? Seriously, *why*? As I had said to Nick, having the alarm out there wasn't going to change our numbers for the year; our go-to products were the doorbell, the new floodlight camera, the subscriptions that were growing by leaps and bounds.

Maybe the mission was driving me crazy. Maybe I had become so subservient to it that I couldn't see straight anymore. I was going so fast and so hard, I had actually put all of us in harm's way.

How arrogant had I gotten? Just because I had made other bold moves that had worked out (along with many that had not), did I really have to do this? Was it a "fuck you" to everybody? Was I trying to prove some bizarre point?

And so the moment we released the alarm, we unleashed a hellstorm of trouble for ourselves, and put the whole endeavor of building a great company devoted to better security on the verge of collapse.

I could argue all I wanted that ADT had screwed us. But when we released the alarm, inviting pushback that I should have known was coming, and that made our already shaky financial situation way more so... frankly, *I* looked like the asshole. Their boot was on our neck and I had put it there.

I had thought right made might. I was no longer sure I was right.

O

On the plane back from Kiev, I considered our position.

We were in an $80 million hole.

We had enough cash left for just a couple of pay periods.

We probably weren't going to close the latest funding round.

We probably couldn't find new investors right before the holidays, with a cloud over our head just when most VC firms have checked out to go skiing or to Hawaii.

We couldn't sell the alarm, and we would be in big trouble if anyone bought one.

We would get bad press right before the holidays.

We had a huge FUCKING lawsuit hanging over our heads.

Back home, my team already knew about the devastating news.

Aside from that, things were cool.

We went into triage mode. I knew what I had to do. Leila knew what she had to do. Others had their tasks, too. First, we had to get our alarms off the shelves of Home Depot and Best Buy and every other retailer and stop the delivery of those en route on trucks and planes. A premise of the injunction was that if ADT were ultimately to prevail in the lawsuit, harm would have been done to them if any of our alarms had been sold. If we didn't make a heroic effort to get every single one off the retail shelves, we could be seen as treating the matter with indifference and contempt and any ruling against us (unlikely and unfair as it might be) would probably be that much more punitive.

Oh, and we also had to remove all the product and stop all the orders *without spooking our partners* into thinking we had done something wrong. I would not use the word "recall," the single scariest word in retail. But what had just happened was now public. "We're pulling back some inventory, that's all," was probably the least heinous euphemism.

Then we drafted an email to customers who had already paid for the Ring Protect.

> Dear Neighbor,
>
> Unfortunately, we are unable to fulfill your pre-order at this time due to a legal dispute with ADT. We believe this case is without merit and will continue to vigorously fight this in court. Rest assured that

we will be releasing a new version of Ring Protect in the coming months. In the meantime, we will be refunding the full amount of your order.

We appreciate your commitment to Ring, and are offering a $50 promo code that you can use toward any Ring Doorbell or Outdoor Camera, as well as an additional $100 off Ring Protect once it is re-released at a later date.

O

Next, we had to alert our manufacturers across Asia and tell them to stop building alarms. Leila, Don, Chien, and I would split up those fun calls.

We also needed money. Don was meeting with every retailer, promising them discounts of 2% if they paid us in 15 days instead of 30; when they balked, he upped the discount to 3%.

I got a call from the owner of a freight company we worked with. "Hey, your factories don't want to turn over their freight," he said. "What's going on? Do you need a line of credit? Do you want us to spot you?"

I got choked up at his compassion and generosity. Also, at the fact that he apparently didn't read the tech business news. But he had no idea the size of the crater we were in. If he knew I needed $80 million, not $1 million, he might not have bothered to call.

Another shoe dropped (by this time there were more than two): The VC that had been the lead for the Series E called to say they were officially out. They were spooked by the injunction. Not a surprise at all but still a gut punch.

Eighty million up in smoke.

It seemed a thousand years ago that I was toggling in my mind between Door #1, the excitement of getting acquired, and Door #2, the excitement of getting more funding and eventually going public. Silly me, I hadn't considered what was behind Door #3... being driven out of business!

Next, I had to calm the troops. Not only did we suddenly *not* have our great new alarm in stores, but to those who had worked so hard to build, promote, and distribute it, it would feel as if the last several months had been a big waste of time. I wrote to the team:

> At Ring, losing a battle certainly does not mean we will lose the war. Ring has faced losses much worse than this and every time we prevail and it makes us stronger.
>
> ADT has picked the wrong group to fight and today I declare Ring War 3.
>
> Protect will still come out, just delayed by a few months and when it does it will be better than ever. We will still become the largest security company in the world.
>
> We will do this by adhering to the same principles we always have:
>
> - Sticking to our mission of "Reducing Crime in Neighborhoods"
>
> - Working hard and coordinated as a team
>
> - Taking care of our neighbors
>
> Thank you as always, I know you are all here for Ring and that together we can win this,
>
> *Jamie*

Jason called from Kiev to say he'd found someone who had made an alarm that we could license for $1 million, a clean start. (The guy wanted a lot more to buy his company outright.) "Should I explore it?" Jason wanted to know.

We owed $80 million, so why the hell not? "Do it," I said. If we were going down, we were going down in flames.

Thankfully, two of our most loyal and important investors, True and Upfront, confirmed that they were still in for this latest round—despite the injunction, despite the fact that our lead investor had pulled out. But they would need a third investor to participate and act as "outside validation" for them to put in money at the new proposed post-money valuation ($1 billion). I would have to scramble to find someone, but at least I knew what I needed to do. *Hey, would you like to invest $40 million or more in a company that's $80 million in debt?*

To top off a fun couple of weeks, Amazon launched the Cloud Cam, "a premium product at a non-premium price." Retailing at under $120. Different from us, but still. Another thing that threatened my focus. I never cared about the competition, never looked to see what they were doing, because if you're doing that, then it means you're behind them looking ahead, not head-down letting others look to you. When you look at others, it means you're in copycat mode or iterating mode. When you have your eyes on the road ahead, it means you're leading. Maybe you glance at the side mirror occasionally, but look too long and you'll crash. The best way to win is by focusing on what's ahead of you, which is innovation.

A truly bad November was getting worse.

We stopped paying our vendors—just stopped. We had enough cash left for two, maybe three payrolls.

We had money coming in from Rings, lots of it, but so much inventory that was stacking up. We had to pay our vendors tens of millions of dollars before the new product would get to the retailers, a massive cash-flow issue. We had a thriving business, an exploding business, but without the immediate money to pay the people to ship the product as we approached the biggest shopping period of the year, we would be a bankrupt business. Simple as that.

I couldn't sleep at night. Two hours was a victory. I was not chill. I was not in balance.

I didn't know for sure, but I suspected that Amazon was sniffing around alternatives to us, something to fill the home-security space. As strong a brand as Ring had become, didn't it make sense for Amazon to look at someplace fresh, a company that did not have a lawsuit hanging over its head? Why would Amazon want the bother?

ADT had their boot on our neck. I couldn't believe it. They were going to cost us everything. Actually, *I* had cost us everything.

By accident—it wasn't my CFO's fault—the whiteboard in our "conference room" had a number scribbled on it, and even though it wasn't explicitly labeled "how much cash we have left for payroll," several Ringers who saw it figured it out. We'd gotten too comfortable with always almost going out of business. It was too normal.

"Why the fuck is this out in public?" I yelled at Mel in private, though I was hardly above yelling at people in public.

"I was tracking it. I'm sorry. I didn't have it turned so people could see."

"Why is it even written down on a whiteboard?"

Erin, Ollie, Erin's mom, Carole, and I flew to Europe, vacation for them, work and growing bitterness for me, with a little vacation on the side. The trip should have been joyous—I was with my family!—but instead it felt like I was mocking reality. I was broke. I'd booked this holiday during my "I think I'm rich" phase, believing I'd have $10 million in the bank by this time, back when I was playing eeny-meeny between the mountaintop of acquisition and the mountaintop of incredible new pre-IPO investment.

No, I take that back. I wasn't broke. Broke means you're at zero. I was minus eighty million. *But hey, let's see Piccadilly Circus and Big Ben!*

We were there because Ring had expanded in the UK market, plus we got some fluff coverage for building the world's most expensive doorbell, the Ring Elite Crown Jewel, gold-encrusted with 2,000 sapphires and the Ring name in diamonds, asking price of $100,000. We'd made 10 of them, to be sold at Selfridges, the legendary department store, with the proceeds going to UK charities that supported rehabilitation for criminal offenders after their release. The British press had already nicknamed the expensive doorbell the "Bling Ring." If the felon we'd helped the Philly police catch for stealing Ring doorbells had gone for one of *these*, I would have understood.

We stayed at the Bulgari Hotel. *This is the last time we're ever doing something this nice*, I thought. I couldn't get a refund on the room. Whenever Erin looked at me, I just smiled. She knew exactly what I was thinking.

We are fucked.

I needed to keep my head down and focus. I had a full slate of media set up. My CMO, Simon, was English, so he had weighed in on how they were different from Americans. He had convinced me not to be in the commercials for the UK market. "The English don't want an American telling them what to buy." When I got downstairs the next morning with Erin, Ollie, and Carole and we took our seats at breakfast before the main press event, it seemed almost as if I was having an out-of-body experience.

Where the hell am I? What are we doing eating here? Why am I here?

I was so frayed that later in the day, when a reporter asked what I thought of Google Nest launching a video doorbell, I broke my own long-standing rule of never mentioning a competitor by name. "It's embarrassing for a company that spends $800 million a year to be just copying me and be behind me the whole time," I said. "I don't think copying is the way to build something great, and they can't copy the future, they can only copy my past."

Jeez. Bitter much? I hate reading that statement. I hate that I made it for so many reasons.

It was Marketing 101. I knew the rules. You don't talk about your competitors, *NOT EVER*. But I was frayed. When they find the black box after plane crashes, the recording of the pilot right before the crash is often a curse. What else is there to do?

At one point, while the four of us walked around London sightseeing and Ollie and Carole were a little ahead, Erin, always unflappable, squeezed my hand.

"It'll be fine," she said. "We'll make it through. We don't need this anyway, right?"

"No," I said. "We do. It's really bad."

She nodded. She squeezed again. "It's fine," she said.

○

One afternoon in late November, I got an alert from my team about a capacitor problem with the Ring Video Doorbell 2, several thousands of which had already been shipped to the QVC warehouse outside of Philadelphia in preparation for our upcoming Black Friday event. It was not a fix that could be dealt with by a firmware update. It would require opening each box, swapping out the bad part for a replacement, and repacking so that the seal did not appear to have been broken.

I gave Dan and Tom, my great reps for the QVC appearance, a heads-up, no need to panic.

"Let's just cancel the airing, Jamie," said Tom.

"No! Totally unnecessary," I said, defaulting to everything's-gonna-be-okay mode when, as the previous weeks had shown, it so clearly was not. "We have almost a week. No problem. We got it."

"How can you fix that in days? Jamie, it's thousands of units. Listen, take care of what you need and we'll just rebook another brand—"

"Tom! Dan! We'll take care of it!" I was almost yelling. We could not get bumped. My company would start reeking of the smell of death. I needed money, momentum, *some* positive news going into the holidays. Mel had been clear that we were just about out of cash, even as we weren't paying vendors we owed, and were haranguing those that owed us to pay faster.

Still, how to fix the problem? I was dead on my feet, maybe coming down with the flu, and it would be tough to sell like a maniac for a 24-hour marathon if I knew that every QVC customer who bought a Black Friday Today's Special Value Ring Video Doorbell 2 for $159, with five easy payments of $32, was about to get shipped a faulty product.

Correcting a capacitor issue in thousands of packaged units on the other side of the country would take effort and cleverness, and Dave Savage, now head of logistics, was on it. He and his team flew to Pennsylvania, took over a section of the QVC warehouse, unpacked each unit, fixed it, then followed QVC's strict protocol for repacking without making the package look used.

Mimi and I flew to QVC for the gig. With less than an hour to go before midnight, when I was scheduled to go before the cameras to kick off Black Friday, Dave called to say they had taken care of every single one.

I was psyched to move lots of doorbells, but I was also getting sicker by the hour. In the QVC studio, I knocked back Red Bulls in the green room while Mimi made pots of lemon tea. To keep from passing out, I had someone get me chewing tobacco, Kodiak, a habit that my buddy Jayson Rully and I had picked up in high school. We called it "The Bear."

When Tom and Dan poked their heads in the green room, they looked nervous. A very *Is he gonna make it?* vibe. I thought about *Shark Tank*, right before the taping, me wondering if our last doorbell would work. This time, it was the ones behind the camera who were freaking.

At midnight, the end of Black Friday and the beginning of our marathon, I went on for an hour, doing demos, taking calls, bantering, smiling, epically upbeat. The wifi in the studio was iffy but I'd learned how to work around it. I really hoped the mountainous display of Rings did not topple over because of my stash behind them. At one point the co-host picked up a Ring box for a close-up, revealing several Red Bulls and a cup of tea.

The scene of me as bartender five years down the road played in my head.

We sold $1 million of product in the first hour.

When the segment ended, I nearly fell down, then staggered back to the hotel for a few hours of tortured sleep. I was back on at eight in the morning, even sicker, then gritted through 8 to 10 more airings, 22 to 60 minutes each, throughout the day and night, a 24-hour event. With each segment, I could see Tom and Dan in the wings along with Mimi, all of them exhorting me with their eyes. *Just keep it going, keep it going. You got this! Don't fall down! Dyson! DYSON!*

I was on live TV for 12 of the 24 hours. At the end, I got to stand in front of the "SOLD OUT" sign. We had moved 140,000 units, for a haul of $22.5 million. I was told it was a QVC one-day record for a single product for that year.

Might as well go out with a bang, right? I thought of Lori Greiner from *Shark Tank*, who did so much business on QVC. She would be impressed by how much product I had moved.

Cold comfort. It didn't matter. I could see no way out of our predicament, but I also knew that curling up in a ball or waving a white flag was not me. Stopping = Death. My world was burning, and the only way to get out was to move. Move ahead. It was so Jekyll and Hyde: When it was just me and the product I'd created and improved with my incredible team, when it was me pitching its value, customers totally getting that it was a doorbell but so much more than a doorbell, what it meant for their lives, their feeling of security and ease... we couldn't be stopped. All the other stuff faded in the background, at least for one day.

Then there was the lawsuit. Amazon pulling its term sheet. Negative press all over. Me just giving up on sleep, period.

As we left the QVC studio, Mimi wondered about the sound clanking inside the backpack slung over my shoulder. I'd taken a six-pack from the green room.

"You took beer from the green room?"

I shrugged.

I splurged on a helicopter for Mimi and me back to New York. The helicopter company took credit cards, so why not? We drank the beers as we floated past the spectacle of Manhattan still lit up and sparkling at one in the morning. When we passed over the Statue of Liberty, I took a picture to help remember it all.

O

The end of November, Leila poked her head in my office.

"ADT called to ask if we want to do binding arbitration in Miami," she said.

"To talk settlement?" I finally asked. It took a moment for me to realize I had enough air to speak because I felt the boot slightly releasing from my neck.

"Yes, to talk settlement."

"Holy shit," I said.

It was the spark of a spark of a positive sign. We had thought they would never be open to settling. It was as if the lawsuit had banished me to rehab for overstepping jerks, and now I had a chance at redemption. "Do you want to settle?" hadn't even been in the realm of possibilities, and now it was.

But I wasn't getting too excited, or even a little excited. Because I expected that the dollar amount they would ask for at the settlement hearing would easily put us out of business. Still, we had to go.

December 7—the date of the attack on Pearl Harbor—was the only day it could be done before the holidays, because of the schedule of the retired judge whom ADT had chosen to preside over the arbitration. If we didn't do it then, we'd have to wait until after the new year, a date that probably would have been too late to save us.

To make the hearing, I had to renege on some commitments I'd made months earlier, including an appearance December 6 on QVC Germany to spur that market, where we'd recently launched. I hate going back on my word but there was no other way. I called Tom and Dan in Pennsylvania to tell them, with deep apologies, that I would not be co-hosting the Ring Today's Special Value in the QVC studio in Düsseldorf. I didn't want to go into detail about why. They told me how disappointed the Germans would be. I said I understood, I was really, really sorry, but I simply couldn't do it. The timing was impossible. I loved that QVC was always done live but this was one time I wished they pre-recorded.

I tried convincing Dan that he could do just as good a job co-hosting, but QVC wouldn't okay it.

No excuses. The words ate at me.

"I can't, guys," I said. "I don't take it lightly but I literally just can't."

I could hear the disappointment in their silence. The Germans really, *really* wanted the Ring founder.

I told them I'd hire a professional TV host to do it, anyone they wanted.

Pull Jamie and we pull the segment and send the stuff back, the Germans had told them.

Are you shitting me? Send product back? That's the last thing we needed.

There were 12 hours between the time the Germany gig ended and the start of the mediation in Miami. I knew enough about planes and the laws of physics that it seemed possible—but highly implausible.

I told them I would do it. I hated breaking a promise almost as much as I hated excuses.

Going down in flames, baby!

O

I landed in Düsseldorf, Germany, as exhausted from a flight as I'd ever felt, took a car to the QVC Germany studio, and right away learned that we were having wifi issues. I couldn't help but flash back to a drenched Mark Dillon trying to get our last DoorBot to work before the *Shark Tank* cameras started rolling. Then one of the Düsseldorf studio staff confessed that the problem wasn't actually a problem but the point: They purposely throttled down the wifi in the studio to mimic that of the weakest houses in Germany because "that's what the German population is used to." The fellow was just being honest, and it wasn't his fault, and I even kind of admired the logic. But I was so close to exploding.

In the intervals between on-air selling, I finally shared with Dan why I had tried to cancel the trip: It seemed likely that I was going to lose everything after I returned to the US. I told him I had chartered a Gulfstream jet to fly home, because it was the only way I could make it on time from Düsseldorf to Miami for the arbitration with ADT. I asked him if he wanted to join me but he said he couldn't. He'd been doing lots of traveling, and had a wife and kids at home. Of course I understood.

I went back on air. We had done such a great job at building the brand in the US—Simon leading the way, with great help from Karni and Mimi—but Germany was proving a harder nut. No one on the team had lived there. No one had a special understanding of the market or culture.

Didn't matter: We met our goal and sold all the product on hand at QVC Germany, a good haul for that market. Pretty much all the profit we made would be eaten up by renting a Gulfstream to fly one person across the Atlantic Ocean. Outside the TV studio, a car waited to race me to the airport. Dan and I shared a big hug, and he wished me good luck and gave me the only smile one could give for such an occasion.

At Düsseldorf Airport just before midnight, I boarded the Gulfstream. It was just me and the crew. I was bummed Dan wasn't there. I could have used the company.

I've always adored planes. When I was a teenager, my Uncle Adam had a place in Virginia and his family often flew a small plane down there. I sometimes got invited, and always asked to sit up front with the pilot. I found the design of some of those machines beautiful—the Pilatus, Citations, and Falcons.

None of those compares to a Gulfstream. This was a Gulfstream IV-SP (for "special performance"). I had always dreamed about flying one day in a Gulfstream.

But not like this. In my fantasy, there are other passengers in the GIV besides me. People I love. And we're heading to Hawaii or Paris. I'm not flying by myself to Miami (sorry, nothing against Miami) for a meeting at which my company is officially getting choked out. In the fantasy, I am not staring at my laptop screen, creating a PowerPoint for my board of directors, outlining the steps we'll need to follow once I fire a thousand people who for years have done nothing but make another dream come true, the one where we make the world a little safer. Sure, in my fantasy I'm probably being served drinks in fine crystalware, as I was right then. But in this case, it was just me guzzling vodka sodas so I could pass out under an expensive soft blanket in a haze of stressed, drunken depression.

Another thought that kept banging in my head: *How do you explain to your team, as you fire them, that you just flew home on a Gulfstream?*

I had become almost incapable of sleeping. Was there a point to what I'd just been through? The bartender scene was now looping.

Thinking back, I'm reminded of a funny line by Mark Cuban. Once, he was asked if he felt dumb for selling a stock at $200 that had then climbed to $230. "It's hard to feel dumb when you're flying around in your Gulfstream," he said.

True, it wasn't my personal Gulfstream, but he was wrong. You *can* feel dumb in a Gulfstream. Right then, I felt very dumb. Maybe the biggest loser ever to fly on a Gulfstream.

Between drinks, I labored over the plan to shut down the company: reducing customer service to 50 people, selling what was left for scraps.

Speeding across the Atlantic, I knew Erin and Ollie would never be disappointed in me for what I had tried to do. But I was disappointed for them, for all the occasions I'd missed, even though I'd done everything to be there for them.

I texted Erin that I would have paid any amount of money to get out of doing this thing in Miami—but, hey, I was about to pay an insane amount of money regardless.

> I know but just enjoy it. You've certainly worked hard enough.

> I just wish I was with you.

> > Would have been fun

> That is the sad part. Alone on a freaking jet from Europe.

> > To go pay ▉▉▉▉▉ in Florida.

> > Yuck

I felt as if I was driving to my own funeral, only at 45,000 feet. Yes, the Gulfstream flies higher than commercial jets.

For the first time in forever, I fell into a deep, full night's sleep.

CHAPTER 15

DONE

———

Leila, a couple of other lawyers, and one of our board members had flown to Miami from LA for the binding arbitration. I was glad for the support, and they absolutely needed to be there as key members of the Ring team. But frankly, I didn't see how it made any difference to the outcome. It felt like a wake, except wakes tend to be more upbeat.

We were in one conference room, the ADT group in another, with the mediator prepared to go back and forth with competing offers and counteroffers. He was a retired, white-haired judge. Because the process was binding, when the day was done, if both parties agreed to the settlement figure and terms, we'd sign it then and there and be done.

For the first half of the day, nothing happened. We didn't understand. *They* had called *us* to settle and, what, they just wanted to see if we'd show up? *Yes, we're here! In the room across the hall! How about an opening bid? Even a crazy one, so we can start negotiating?* Sure, we didn't have the money to pay them but at least we could pretend, right?

Even a terrible offer would have made sense. Suddenly Mr. Wonderful's pitch seemed absolutely reasonable.

Maybe they were arguing among themselves. I really didn't know. None of us in our room did. The whole thing seemed like some childish exercise.

So many things about this killed me, and the part about how we had legally used components and code in the alarm that they no longer owned was not even near the top of the list. The facts seemed no longer to matter. They were going to squeeze us out of business, whatever it took.

Erin texted:

> You think you are coming home tonight

>> Doubtful

> Does that mean its good or bad?

>> Bad

> Dang

>> Yeah

I had bought two plane tickets, both leaving Miami the following day, December 8. One flight was to LA, where I would fire a thousand people and refine the PowerPoint to outline how we could sell the company for a small price and get some—not all—the money back for my investors. I would get zero in equity for my work of the past six years.

The other flight was to Montana. If I somehow actually walked away from the day with my company still standing, I would head to Big Sky Country to go skiing with my friend Scott.

After lunch, Leila, the others, and I trudged back into our airless conference room for another four hours of performative inertia.

The judge walked in.

"I don't know what happened," he said, "but you have an offer."

He showed me the dollar figure.

I showed it to Leila and the others. We all looked at each other, as dumbfounded as we were encouraged. It wasn't catastrophic. Of course, we would never accept the first offer. Even in my darkest hour, I had to follow good business principles. But we could work with this.

Very weird. Had they given themselves the morning to see if we would offer something first, which was never going to happen? Had something spooked them? Had eating lunch spiked their blood sugar?

Nothing, nothing... and all of a sudden, something.

"I think we might walk out of here at the end of the day," said a cautiously hopeful Leila.

Bizarre.

My team and I sent back a counteroffer, which we felt was fair and could possibly get done. It didn't; not right then. For the rest of the afternoon, we went back and forth, the mediating judge earning his fee. Two, three, four times. I gave Erin the blow-by-blow via text.

Miami was dark outside, dinnertime. Both sides had finally agreed on a settlement price. Our side had been tagged, a hard right to the chin. We were wobbly. But this agreement would allow us to stay standing.

Before I signed the document, I shuddered. There had to be another shoe waiting to drop, somewhere. There always is. I paused pen over paper. "How can they get out of this?" I asked Leila.

"They can't," she said calmly.

"There must be a way."

"There isn't."

"There *must* be."

"There isn't, Jamie. It's binding."

"Could they—?"

"No, they can't."

"What about—?"

Leila, who had probably never told someone to shut up her entire life, now did.

I believed her, finally.

I texted Erin.

> settled

Hallelujah!
Can you start selling it now
The alarm not the company

> Both

Well congrats this seems Like a positive step

> beyond

Now you can enjoy your ski trip

> Oh fuck yes. Buying that condo this weekend

Slow your roll turbo

Can you hire back all those people you
had to get rid of?

> We never got rid of them. They are
> going to go nuts tomorrow

The huge question—*Why?*—continued to eat at me, but I had to let it go.

I'll never know why ADT made us wait, and then why they settled. It didn't matter and it doesn't. They could have gotten way more from us than they did. We would have cried uncle at a higher price, for worse terms.

Thank you, ADT. The whole Ring team thanks you. War is over.

○

In the car to dinner, I called Nick at Amazon. I was just hours removed from believing we were finished and that I would have to fire a thousand human beings, right before the holidays.

"It's all settled," I told Nick. "I cleaned up my mess."

"We'll get back to you—but this is really good news." He had turned from colleague back to friend. "We'll be sending you a term sheet."

Within days, Upfront and True did a bridge round to help us cover our debts. Knowing the lawsuit was behind us, they trusted that the new money going into Ring was a good investment.

Our sales were way above projections. I had told Nick we might do $180 million for the quarter. Amazon had surely been tracking our progress, and since our volume on their site reflected how we were growing overall, they knew we were exploding. Throughout our history, during all of our down times, even during this debacle, we had kept our heads down and kept building, selling, working, while weathering the storm. We were a young, hot company that was ripping. And now we no longer had a dark cloud hovering above us.

The journey was hardly over—I didn't know the terms of the Amazon offer; there would be weeks, if not months, of due diligence—but I knew the option I wanted to take. I didn't have the energy to go through another round of funding, even if I was sure we could get our lead investor back, or if not, then another investor more than happy to lead. I wanted to go with Amazon because they had proven to me that they were like us, mission-driven and missionary-driven, with a founder who cared first and last about the value to the customer, customer experience, constantly getting better, never resting. With Amazon's balance sheet, there was so much we could do to improve safety in homes and neighborhoods, sooner than later. The mission was everything. Ring and I would share in that with Amazon.

That night in Miami, I got miss-my-flight-home drunk. But the flight I missed the next morning was the one headed to Los Angeles. I didn't have to fire anyone. I didn't have to plot out the fire sale of Ring.

I woke up later in the morning, hung over, and raced to the airport to catch my other flight, to Montana.

O

After I returned from skiing, I received the term sheet from Amazon.

Many of the people on my board were against selling. By doing so, I was almost certainly leaving money on the table, so they would be, too. But I also knew what I had just been through—me, not them. My Ring team, not them. I had come within a hair of spoiling the holidays and livelihoods and short-term futures of a thousand people. My dream had come so close to being dashed. Our mission, which I cared about more than the company, would have been cut short. After what I had just experienced and what we'd just barely survived, I was not about to gamble.

Ollie could sense my excitement and apprehension. "But what happens if you don't sell to Amazon?" he asked Erin and me over dinner one night.

"If that happens," I assured him, and myself, "then we keep going. That's what we do."

I spent New Year's Eve day at my house, negotiating on video-conference with the board, as we went back and forth with the Amazon offer. One member would say, "Okay, I agree, let's go with Amazon" and someone else would pipe up and be sharply against it. We either had to sign it or not sign it. We talked it out, like jurors deciding the fate of the accused, and I'd finally win over a no vote... only for another board member, who'd been quiet but seemingly in favor of the Amazon option, to say, "You know, So-and-So has a point..." And it would start all over again.

After hours of this, I'd had enough. As I bolted for the front door to go for a run in the woods, Erin came inside from walking the dogs.

"I'm done!" I called out loudly enough for the board members, still on videoconference, to hear. "Tomorrow is the start of 2018!" I half-yelled. "I'm working for Amazon with or without you! You tell me how you want it to go!"

I headed out for a run. Straight uphill would have been my preference.

Later that afternoon, the board called back.

"We voted," they said. "It's unanimous. Congratulations, Jamie. Amazing job. Go sign the term sheet."

In the evening, Erin, Ollie, and I walked a couple of houses over for a party at our neighbors', Gloria and Dan. I scanned the room—so many great friends talking, eating, drinking, smiling. I wanted to tell anyone I could that I had just sold my company to Amazon.

I couldn't. I would have to keep it secret for months.

Sitting on the living room couch, taking in the scene, I thought: *You were right, Mark Cuban.* DoorBot did *not* grow into a company worth $80-90 million. More like ten times that. Then go even higher.

I tried to keep my eyes open but fell asleep on the couch. I woke up right before the stroke of midnight, December 31, 2017. I looked next to me at Erin and Ollie, my two favorite people in the world. I knew what I had to do.

I draped my arm over Ollie's shoulder, lifted a glass of Champagne, and locked eyes with Erin.

"Let's celebrate," I said.

POST-MONEY

In the end, we agreed to be bought by Amazon for $1.15 billion. That makes us a unicorn. At the time of the sale, Ring was Amazon's second-largest acquisition, after Whole Foods.

The deal officially closed on April 12, 2018, at one second after midnight.

On *Shark Tank*, Mark Cuban said he needed to believe that our company could reach $80 to $90 million in value but he didn't "see that progression, and for that reason, I'm out." He had only an hour to get to know me so I give him a little leeway on this one, but he did miss by more than 10x. Today, by my calculations—based on the number of homes with our products—Ring is the biggest home-security company in the world by far. The Ring Video Camera is probably the bestselling consumer camera in the world.

A stupid little doorbell company.

I broke down in tears, several times, during the months of negotiations, due diligence, and closing with Amazon, as if a dam had burst. (Of course, I broke down in tears many times in the years leading up to that, too.) I was exhausted. I got shingles. The pressure, the descent into deeper and deeper debt, the perceived threats at every turn, the knife's edge we walked on at the very end: this side, the possibility of losing everything; that side, a glorious exit. Looking back, it felt as if we were *always* going out of business.

Our "neighbors" loved our products and the company—still do—in just the way Mimi had said consumers can identify with inanimate things. Many of our product launches were riddled with glitches, so how had we continually survived the negative coverage and reviews we got from the media and users? Marketing, branding, the mission. In numerous surveys of consumers, Ring consistently scores very high for both recognition and popularity. There's little room to mistake what we do or who we are.

From the time Diego had to curse me out for not focusing, and I understood that this was more than a doorbell, I had put all my chips in, even though I keep insisting I'm no gambler. But I'm not so full of myself to believe that it had to be this way. It easily could have ended very differently.

One of the ironies: You work and work and work, every day, maybe for seven years or more, to give yourself enough lottery tickets for the best chance to win... yet in just one day, one hour, your dream can turn into reality or a living nightmare. How can it be that such a thin line separates failure and success? Or is it all about our *perception* of failure and success? (Nope. I know what it's like to see a negative balance of $80 million, and that's no perception.)

I learned later that when Amazon was considering whether to participate in the Series E round or buy Ring, Jeff Bezos had emailed Dave Limp (then senior vice president of devices and services), Nick Komorous, and others. "I'd buy it right now," he wrote. Not quite four hours later, he expanded on what he saw as the benefits of the deal to his company: "To be clear, my view here is that we're buying market position—not technology. And that market position and momentum [are] very valuable."

Jeff eventually came to visit our operation at 1523, a true dump. I gave him the tour: where we had done customer service in the early days and had to move shipping outdoors for business hours; the racks where sometimes doorbells would randomly *DING! DONG! DING!* at night and we didn't know which boxes to open; my office with the surfboard table. I showed him the two holes I'd punched in the wall. "I was upset that day," I said, as if that wasn't apparent from the ugly little craters.

"I love it!" said Jeff. Out in the parking lot, he told me, "Jamie, you need to keep the hole in the wall. You need to buy this building. It's special." He didn't mean, *Hey, now you're rich, buy a building*. He meant, *This place is spiritually special. It has your history in it*. I did try to buy the building, numerous times. Sadly, the owner will never sell it.

Ring holds hundreds of patents around products and processes related to reducing crime and increasing safety and security, in homes and in neighborhoods. My team will never stop innovating. They're brilliant engineers and problem-solvers. It helps that I was not a classically trained engineer, because I continue to not know what I can't do, which can be a real help. (Or, sometimes, a real annoyance to engineers who *do* know what I can't do.)

I knew we had made it, by a number of metrics. But nothing told me we had become a household name more than a line by Tom Brady, the 199th pick in the 2000 NFL draft, who joked at his 2024 Netflix roast about his former coach: "Now that I'm retired, my favorite ring is the camera that caught Coach Belichick slinking out of that poor girl's house a few months ago."

To this day, it doesn't seem like the doorbell is a real invention to me, because it was so damn obvious. I think of the role that pure chance and timing played in my own odyssey. Everything connected with *Shark Tank*. The incredible people who answered my Craigslist ads, open for a new adventure. Sir Richard finding out about us the way he did.

Yet it's still all about putting your head down and doing the work.

I still don't gamble in business, but I'm part of an entrepreneurs poker game. I've been given the nickname "Two Hands" by Chamath Palihapitiya of the *All-In* podcast, not because I'm aggressive but because I have a habit (which I didn't realize until it was pointed out to me) of pushing my chips in with both hands, when one would suffice. I'm getting better at reading people.

Could I have kept the company independent and maybe taken it public? Josh Roth says "there's zero chance we couldn't have gotten the money." He might be right. Show me a startup company that's successful that *isn't* underwater for a long time, including Amazon. We were a hardware company with a subscription model. The closest one at the time that did go public, with a $40 billion valuation, was Peloton. And our business was at least as good as theirs.

But Josh—and all my friends who tell me not to linger over the choice I made—knows that no good comes from replaying the final dance, since there were many compelling reasons for selling. Who knows what could have gotten in the way if we'd held on to go public? Would Covid have changed the trajectory of the business? (Actually, it did: Because people spent more time in their homes, Ring's already booming business boomed even more.) It's easy to say we could have sold the company for 10x more. But what if I had just physically collapsed? What if there had been an earthquake? Saving jobs was the most important thing to me: done. And having a supportive, well-funded acquiring partner who could help us get more doorbells and other security products on more homes, and bring a greater sense of safety to more people: done. End of story.[11]

And why *did* ADT suddenly, thankfully turn around after lunch at the going-nowhere arbitration meeting and finally show a willingness to settle? Could it be that they figured we were on a path to releasing another alarm (we were; the one we'd licensed in Ukraine), and they figured they needed to get *something* out of this fight? Maybe. Could it be that they were preparing to go public? Maybe. I'll never really know, until ADT writes *their* book. I know only our side.

As for mistakes I made:

---

11 Okay, not end of story. Every so often I still keep myself up at night, wondering what-if.

Had I hired more senior people, maybe it would have helped us scale even faster and things would have been less crazy. Certainly it would have made this book less fun. Because I didn't hire many established people, I was never able to raise enough money in each round to get us out of *constant* startup mode. Don't get me wrong: I still feel youth and inexperience are great—they helped make Ring the incredible company it is today. But I might have saved myself many anxious moments had I been willing to let go and bring in some people who knew stuff that I didn't. But I wouldn't slow down (Asshole!), which was both a plus and a minus.

Speaking of assholes: Maybe I shouldn't have acted like one to the ADT executive on the phone. Maybe—no, definitely—I should have delayed the release of the Ring Protect alarm until after Amazon acquired us or we had closed our Series E round. That was arrogant and dumb.

If I came off to people as if I thought we were the best in the business, or we had practically invented the industry: I was able to believe that because I kept my head down, my team kept its head down, and we always made it about our customers, our neighbors, not the competition. And we made it about us, not the competition. What *we* were capable of. We were like golfers playing against the course, rather than the rest of the field.

So... where is everyone now?

Our investors:

All of them made their money back and more.

Richard Branson: My chance elevator meeting in a Brussels hotel with Sir Richard helped inspire the "Elevator of Dreams" in one of his London hotels, where budding entrepreneurs can pitch their ideas to a global audience and see who bites.

Ollie and I traveled to the Mojave Desert to see a Virgin Galactic rocket launch.

Latif Peracha: Sir Richard's right-hand man was essential in bringing in the investment. He wanted to put his own money in, too, but was told no by his boss. Sorry, Latif. That didn't age well.

Josh Kopelman: Right after the sale he bought a bunch of Rings, had them woven into a crown, and gave it to me with the inscription "The Doorbell King." There's no one whose friendship and guidance have meant more to me throughout this journey—and he was a friend and supporter before it all started.

Steve Spinelli: In 2019, Babson College made the brilliant move of naming Steve, my favorite college professor, its president. Five years later, the *Wall Street Journal* rated Babson, always punching above its weight, one of the top three colleges in the US, behind Princeton and ahead of Stanford. Steve emailed his fellow presidents at both schools to say, "Hey! It's the three of us! Isn't that great?" He has yet to hear back from either one.

Scott Marlette: He remains one of my closest friends and most thoughtful confidants, and I'm grateful for his continued pragmatism, a necessary counter to my borderline lunacy.

Diego Berdakin: He was so crucial in collaborating on early specs for home security and served as a constant idea machine (and investor, and noodger of other investors) as the doorbell evolved into DoorBot and then Ring.

I reneged on the bet we made, that if we topped $30 million in sales the year the Ring Video Doorbell came out, I would give him my Land Rover. He won the bet easily; we blew by that number. But I'd grown attached to the car. I guess I think Jeff Bezos is right: When your history is wrapped up in certain places or things, you should try to hold on to them. I wrote Diego a check for the cost of the vehicle so he could buy his own.

And one more thing, my friend: You told me not to pay more than $100,000 for the name Ring.com. Now that you've read this, you know

I paid much more. You're the smartest person I know, but on this one I think I was right.

Hamet Watt: The now former partner at Upfront, who pointed out that I should call my company Ring since I kept saying "ring... ring... ring," recently started an innovation lab called Share, which funds startups. He called one day and said, "Hey, I'm thinking about doing a hardware company."

My pause should have been answer enough.

"You know how hard those are, right?" I asked.

"Well, yeah."

"Okay... what is it?"

"The toothbrush."

It's actually a great idea—his team uses computer vision to do real-time scans and detailed analysis of the teeth, gums, tongue, and soft tissue. He showed me renderings and before you know it, I was in.

Kevin Dunlap: He's no longer sorry that a doorbell salesman chased him down to discuss business at a rooftop cocktail party. Kevin's humility—and practical smarts—to let Sir Richard lead a round that Kevin and his company had initiated was instrumental in helping Ring achieve another level of exposure and popularity. I'm lucky to call Kevin my friend.

Adam D'Augelli: He was more therapist than board member. Or maybe that's what a board member should be. I am forever grateful for all of his unwavering support.

Mark Suster: I will never forget his coming to the office, when we were completely upside-down, and telling me he was still in the round. Mark's also a big champion of our failed Ring Explorers idea: Yeah, it didn't work for us, right then, but he recommends the idea behind it—find ways to get your customers to become product champions—to any open-minded entrepreneur he meets.

Chris Fralic: One of my most stalwart supporters, despite lots of evidence that he was nuts to keep fighting for us. Now that Ring makes such great doorbells, Chris is able to walk his dog around the neighborhood in peace. (I sent him a dozen of the working ones to give to his neighbors.)

Sky Dayton: Every entrepreneur wants to hear that a VC has agreed to give them money. But in a way, there's something even more special when a VC says, "I believe in you and I'm going to invest. Even if I don't think it's the right idea." And it's beyond special when that VC is a legend. Thank you, Sky, for everything.

Chamath Palihapitiya: He's brilliant; he runs exciting companies; he co-hosts *All-In*, a popular podcast; and he's included here not because he invested—he did not—but because he so freely admits that he was an idiot for passing on the chance. (Though it's not as if he makes these sorts of mistakes often.) He's not-so-slowly making back the money he could have made from Ring in poker. From me.

Nas: The great rapper made a wise investment. I don't know if the return topped what he makes on his albums or tours, but I would be surprised if Ring was not his biggest hit. If it isn't, then he is really, *really* rich.

Those from *Shark Tank*:

If not for *Shark Tank*, Ring would have never made it. The support it gave us was almost incalculable. I am forever grateful for everyone that makes that show happen. It enabled my American Dream to come true.

I was the first entrepreneur ever invited back as a Shark—which seems funny, since I wasn't even funded by them. To be fair, though, I'm pretty sure Ring is the still the biggest company ever to have been on the show. I appeared again on Season 10, Episode 15. I entered the same way I had on my first appearance: calling out to the Sharks from the other side of the big door. Kevin "Mr. Wonderful" O'Leary greeted me with,

"Well, well, well... I was the only one who believed in you. You deserve your seat here."

As a Shark on that episode, I bid to invest in Moink, a company pitched by Lucinda Cramsey, a dynamic eighth-generation farmer from La Belle, Missouri, whose energy reminded me of my own. The company ships boxes of pasture-raised meat ("Moink" = Moo + Oink). Right after the episode aired, Moink saw a big spike in sales, just as I had with DoorBot. (Thank you, *Shark Tank*!) In 2024, the company did almost $20 million in sales.

Mark Cuban: In 2023, he sold his majority interest in the Dallas Mavericks, with the team valued at $3.5 billion (he still holds almost 28%). I'm pretty sure Mark's feeling okay in his Gulfstream.

Kevin "Mr. Wonderful" O'Leary: In a CNBC interview in 2018, he was asked if he regretted not getting a yes from me on his DoorBot offer. "Yeah, I'd like to have had another hundred million," he said.

Kevin and I became friends. Every July, we co-host a lunch for entrepreneurs.

Robert Herjavec: I was invited to his 60th birthday party. I sat with Lori Greiner, Daymond John, and other Sharks. Robert and I have remained close and share an unnaturally deep love for our dogs, Belgian Malinois.

Our Shark Tank producer: She was my cheerleader, my champion, and did her damnedest to keep me loose, even though I froze up the moment I got out there.

It all worked out in the end.

My partners, many of whom are friends (and friends, many of whom are partners):

Rami Rostami: Though we went our separate ways on Ring (DoorBot at the time), we remain close. And Rami just became a grandfather. Congratulations!

QVC: They bought their archrival, HSN, in 2018. My good friends at Media One Products, Tom Czar and Dan Forde, my contacts to QVC, like to say that the record-breaking haul we did on Black Friday 2017—$22.5 million of product in 24 hours—ultimately helped seal the deal with Amazon.

On April 11, 2018, after all the due diligence was done and all the paperwork had been submitted, I had to wait until midnight to see if the Federal Trade Commission flagged anything that could potentially hold up or derail the deal. I was in Philadelphia for dinner with Josh Kopelman and called Tom and Dan to meet us, and the four of us ate, drank, talked, laughed—and, my eye on the clock, toasted every time another half hour passed with no word. I was in touch with Erin the whole time (of course), and we all cheered and cried when midnight finally came.

Karni Baghdikian: If there's one thing Karni is better at than making great commercials, it's rugby. He was not just an All-American at Babson, but through his and others' support of the program, our tiny little Division III alma mater stunned the college-rugby universe by winning the National Small College Rugby championship in 2023, against some Division I schools. Double F you, Princetons and Stanfords of the world!

Chien Lin: A Californian by way of New Jersey and a Taiwanese American by way of Texas found a way to collaborate, help each other thrive, and become close friends. And no, I will never look at a plate of hairy crab ever again.

Nick Komorous: I ended up buying that condo in Montana after the Amazon deal closed. Nick joined me there and we raised a glass.

Shaq: Shaq is Shaq. Maybe it sounds silly or starry-eyed, but one of the most fun outcomes of this whole odyssey is that I have Shaq on speed-dial. When I call, he not only usually answers by the second ring, but he always greets me with, "What's up, boss?"

My Ring team:

I am so relieved and thankful that I did not have to fire any of the people I was thinking about on the Gulfstream flight from Düsseldorf to Miami.

Mark Dillon: We parted on good terms when he left Ring in 2014. He wanted to be in New York. Even though we like to believe that remote work is as effective as in-house, it's not. (It's also nowhere near as good at developing camaraderie, which was so key to our success.) The cross-country commute got wearying pretty fast. But Mark saved my ass numerous times. If that fourth and last doorbell hadn't worked, you wouldn't be reading this sentence right now.

(Had we been on *Shark Tank* a couple of seasons later, after the show moved to a different set on the Sony lot, Stage 26, we probably would have had equally terrible connectivity issues. Right beneath the titanium plates on the floor was the Batcave that Christopher Nolan had built for the Batman series.)

August Cziment: My jack-of-all-trades wore a suit to his wedding. Maybe he didn't burn the others after all.

Giving him the wristwatch my dad had given me meant as much to me as it did to August. Sometimes, material things with a history *should* be given away.

John Modestine: We brought him west of the Mississippi, and he's now gone much farther than that, though I still think of him as that kid from Phil U. No matter how hard he tries to convince me he's a designer, not an engineer, his great work says he's both. He's still at Ring, and at this point I fire him only every other year. (I've calmed down a lot in my later years.)

Dave Savage: Years ago, when he started, he was sleeping on John Modestine's couch and looking after boxes and tape for me at $10 an hour. Today, as director of supply chain for Ring, he anticipates a million things around customer need and product availability, so there's always the right amount of product where and when it's needed. He reports up to the CFO of Amazon and excels at a job that Stanford PhDs couldn't come close to doing as well. It's 12-dimensional chess. I have no idea how he does it.

Yassi Yarger: Employee #8 just kept moving up—from public relations manager (back when she *was* the PR department) to director of communications to head of PR to head of global PR for Ring *and* Blink to PR lead for Ring, Blink, Amazon Sidewalk, and Amazon Key. I don't know that there's anything I take more pleasure in than finding great people who just keep rising and rising.

Josh Roth: We quite literally drew up the idea for Unsubscribe, our first business together, on a napkin. We love each other like brothers and would do anything for the other, at any moment. Also, we're like water and oil, and don't know if we could ever work together again. Then again, I think we said that after the last time we worked together, post-Unsubscribe and pre-Ring.

Leila Rouhi: She's now vice president, trust and privacy, for Amazon devices. The title perfectly captures both her capability and her integrity. I still feel bad that I didn't hold off releasing our alarm. Because of me, her vacation in Iceland was ruined.

Mimi Swain: As chief commercial officer at Ring, she oversees everything from marketing to sales to revenue for a multibillion-dollar business. When she joined us, she was a few years older than all the "puppies" at Ring. I don't know where they—or I—would be without her steady hand, wealth of experience, and joy for the challenge.

Don Hicks: One of the world's greatest salespeople has gone on to start a vineyard, among other business ventures. He and my friend Zach both seem to have the YOLO gene, a knack for really living.

Mel Tang: I finally got a CFO who stayed; his expertise and demeanor saw us through the Amazon acquisition. There was a *lot* to work on during the months before the sale was finalized; Mel and his team made our books look actually professional. He has gone on to great success as a venture capitalist focusing on (*why, Mel, why?*) hardware.

Micah Stone: I took great pride in watching this smart, confident, almost savant-like kid buy commercial time and play the system like a wily veteran. I expect we'll hear more of him.

Jason Mitura: He and I enjoy talking about devices more than anyone else in our circle. Maybe no one I know is wired as weirdly as I am (and just weirdly in general). That could explain why Ollie thinks he's the GOAT. Jason has been a hugely valuable brother in arms, not just in Ukraine but back home, too.

Simon Cassels: Hiring Simon was almost the final straw for Erin. His sons were buddies with Ollie, his wife was one of her closest friends, and I threatened it all when I hired him. I couldn't fuck that one up because there was too much at stake. Thankfully, Simon is not just one of the most creative people (Spectrum is lucky to have him as their chief creative officer) but one of the easiest to work with. Our families—big sigh of relief—remain close.

Mark Siminoff: When he left Ring, he went straight back to his shitty business (compostable diapers). Then he sold that company, and he now travels the country in his very modified camper pickup. You can't take the burning man out of Burning Man.

The Ukrainian team: One of the things I am most proud of is what Jason and the team built in Kiev. The office there had such an incredible culture and created so many of the impactful technologies our neighbors love. When the war started in February of 2022, I felt like we went from running a business to running the Red Cross. The support the Amazon team gave us was incredible. Because you could enter Ukraine but not leave, we bought vans in Poland for one-way trips: We would fill each with food, medicine, and other critical supplies; someone would drive them across the border; our friends over there would unload the vans; and there the vehicles remained. The team in Ukraine appreciated greatly what was being delivered but had a request: Could we start including dog food, for all the pets left behind by families that had left the country in a hurry? It's good that I'm writing this: I am incapable of talking about this episode without sobbing. It's one of the most human things I have ever heard, next to one of the most inhuman.

As for 1523 and the culture I fostered there: Not long ago I was among several people invited to speak at a "founder camp" about work culture. The entrepreneur who preceded me, Amy Errett—founder of Madison Reed, a hair-care and hair-color products company—took the stage and talked movingly and persuasively about her company's culture, which she boiled down to one word: love. Then I took the stage and talked about Ring's culture, which I boiled down to one word: war. We were both right. There is no single way to be successful or to inspire a group of people. Just make sure the culture you choose is yours.

My family, close friends—and some others:

My mother: Several years ago, she was diagnosed with pancreatic cancer, which can be a death sentence for far too many people. When I sold Ring, I donated money to Dr. Santosh Kesari for moon-shot cancer treatments. He had been working for years on curing glioblastoma and other cancers, and had seen my father when he was sick. My mother received a vaccine that Dr. Kesari and his team developed—and it worked. In the years since, there has been no recurrence. You might say that she was saved by my father.

My mother undoubtedly gave me many of the qualities that have helped me in life, particularly in business: contrarian, tough as nails, swears like a truck driver, and—most important of all—she's unstoppable.

Often we would go at it, probably because of those similarities. I wasn't always sure how she felt about my path in life. But right after we sold Ring, my mother went to get her hair done at the same place she's gone every week for decades. According to her, she sat down in the chair and said to the hairdresser, "Well, my son just sold his company for a billion dollars. With a *B*."

My brother, John: After we lost our dad, John fell seamlessly into the role of my biggest supporter and trusted ally. He listens to my crazy ideas, revels in my wild adventures, and will always be my first investor. He's the constant to my chaos. I couldn't be more grateful he's my brother.

Olga Iglesias: Erin and I could not have worked full-time and I could not have built Ring without her. She was always there for Ollie and us and always will be. And if you know of anyone else who has ever made what turned out to be a quarter-million-dollar taco, feel free to contact me at j@ring.com.

Zach Vella: He continues to go big, dream big, build big, and live big (now, with the money). We even share reality TV fame: He spent a few seasons on *Million Dollar Listing*. Friends for over 45 years but, truthfully, we're more like family.

My garage: My temple of invention burned to the ground in the horrific Southern California wildfires in January 2025. The physical space is gone but its spirit is alive and well as we launch new and better products from my somewhat bigger, better-funded garage today.

National Package Protection Day: It never really became a thing. We're still trying to make it a thing.

Boston Children's Hospital: There are no words that can accurately capture what Erin, Ollie, and I think of them. The team there, including Dr. Gerard Berry, are the Ring of children's hospitals.

Dr. Santosh Kesari: He and his team have been doing truly paradigm-busting work on glioblastoma, and the research we have funded has resulted in people with this still-incurable disease living years longer than they're "supposed" to. *There's* a man with a mission.

The most important two:

In the early years of Ring, Ollie, Erin, and I would sometimes get in the car at night and drive around the neighborhood, counting how many houses had our doorbell. I miss that.

Oliver delights everyone he meets, he thrives in sports and school (okay, math, not so much), and he will never be topped as my favorite traveling and adventure companion. (Sorry, Don.) However I viewed the concepts of "purpose" and "mission" before I became a dad, they were redefined the instant Ollie came into the world.

If not for Erin, there would be no Ring—the product, the company, the saga, the ultimate success, the book about it. The crime reduction. The ability of its founder to get through it. (Great. Now *I'm* talking in third person.) I can give all the counsel in the world about what works in business, but the single best advice I have is this: *Find your Erin.* It's the equivalent of having 10 million lottery tickets.

○

I was at Amazon for five years after the acquisition, and we grew Ring immensely, into what many consider, based on the number of households served, the largest home-security company in the world. Much as I loved it, I'd been running too hard for too long, so I left for a year and a half and tried my hand at another company, then returned to Amazon, thankfully, in the spring of 2025. Later in the year, I had my Steve Jobs moment, launching a slew of innovative, transformative products and services, including Search Party, another way to improve safety and security—this time, for pets. It will, we hope, help find many of the thousands of dogs and cats who go missing in the US each year.

Part of me thinks we're still not successful. I certainly believe the job is not done. With AI, there is so much we can do with the foundation Ring has built. In some ways, it feels like we're back at 1523, with unlimited opportunity and impact ahead.

I have not quite lived up to that text I sent to Erin from Ukraine in late 2017, the one where I mused that I needed to be more balanced and chill, particularly given what we had achieved. I still have a ways to go on balancing and chilling, but I'm getting there. Even if we had *not* succeeded, and instead simply failed spectacularly, I would sleep better at night knowing we had done *everything*. We put it out there. We took lots and lots of swings, missed most of the time, but made sharp contact many times, too. In the end, I'm glad to say that we were in the arena, our faces marred by dust and sweat and blood, spending ourselves in a worthy cause.

Scan to view photos from the early days of Ring.
thedingdongbook.com/gallery

ACKNOWLEDGMENTS

This book would not be what it is without the help of a number of individuals, not all of them human.

A very big thank you to Anthony Mattero, at Creative Artists Agency, for helping this book come into being.

For their contributions, thank you to Priscilla Cardona, my chief of staff, for keeping the trains running on time; Meredith Chiricosta and Big Fish PR for helping to get my name out there the past 20 years; Myles McDonnell and Robert Walsh, for careful reading and improvements; Chip Simmons and Rich Holtz for useful suggestions; Molly Seabrook for design; and Michelle Weiner at CAA for advice.

Thank you to Karni "Scorsese" Baghdikian, for his usual masterful work on the audiobook.

Thank you to my menagerie, present and past—Biscuit and Pancake (dogs), Chicken and Waffles (miniature donkeys), and the late, great Short Rib (dog)—for helping to keep me sane.

I want to thank the people—all 664 of them—of my new second hometown of La Belle, Missouri. We're going to tell a memorable story there, too.

To Andrew Postman, my partner in telling the story of Ring: You didn't realize that you weren't just writing about F5, but going through it yourself, with me. We should not have been able to launch this when we did—yet we did, hopefully without breaking too much. Thanks for the amazing work, collaboration, and being aligned on the mission—and for the friendship.

And for all those I forgot to name: Thank you.